Lecture Notes in Networks and Systems

Volume 382

The series "Lecture Notes in Networks and Systems" publishes the latest developments in Networks and Systems—quickly, informally and with high quality. Original research reported in proceedings and post-proceedings represents the core of LNNS.

Volumes published in LNNS embrace all aspects and subfields of, as well as new challenges in, Networks and Systems.

The series contains proceedings and edited volumes in systems and networks, spanning the areas of Cyber-Physical Systems, Autonomous Systems, Sensor Networks, Control Systems, Energy Systems, Automotive Systems, Biological Systems, Vehicular Networking and Connected Vehicles, Aerospace Systems, Automation, Manufacturing, Smart Grids, Nonlinear Systems, Power Systems, Robotics, Social Systems, Economic Systems and other. Of particular value to both the contributors and the readership are the short publication timeframe and the world-wide distribution and exposure which enable both a wide and rapid dissemination of research output.

The series covers the theory, applications, and perspectives on the state of the art and future developments relevant to systems and networks, decision making, control, complex processes and related areas, as embedded in the fields of interdisciplinary and applied sciences, engineering, computer science, physics, economics, social, and life sciences, as well as the paradigms and methodologies behind them.

Indexed by SCOPUS, INSPEC, WTI Frankfurt eG, zbMATH, SCImago.

All books published in the series are submitted for consideration in Web of Science.

For proposals from Asia please contact Aninda Bose (aninda.bose@springer.com).

More information about this series at https://link.springer.com/bookseries/15179

Alexiei Dingli · Alexander Pfeiffer ·
Alesha Serada · Mark Bugeja · Stephen Bezzina
Editors

Disruptive Technologies in Media, Arts and Design

A Collection of Innovative Research Case-Studies that Explore the Use of Artificial Intelligence and Blockchain within the Media, Arts and Design Sector

Springer

Editors
Alexiei Dingli
Department of Artificial Intelligence
Faculty of ICT
University of Malta
Msida, Malta

Alesha Serada
School of Marketing and Communication
University of Vaasa
Vaasa, Finland

Stephen Bezzina
Department of Artificial Intelligence
Faculty of ICT
University of Malta
Msida, Malta

Alexander Pfeiffer
Emergent Technologies Lab
Danube-University Krems
Vienna, Austria

Mark Bugeja
Department of Artificial Intelligence
Institute of Climate Change and Sustainable Development
University of Malta
Msida, Malta

ISSN 2367-3370 ISSN 2367-3389 (electronic)
Lecture Notes in Networks and Systems
ISBN 978-3-030-93779-9 ISBN 978-3-030-93780-5 (eBook)
https://doi.org/10.1007/978-3-030-93780-5

This Springer imprint is published by the registered company Springer Nature Switzerland AG
The registered company address is: Gewerbestrasse 11, 6330 Cham, Switzerland

Preface

Disruption of the already established patterns and designs has been the major creative impulse behind many technological breakthroughs today. This anthology introduces a selected number of the latest artificial intelligence and blockchain use cases applied to the media, arts and design sector. The inspiration behind this book originated from the annual Media, Arts and Design Conference that started back in 2019, bringing together the transcontinental community of researchers, educators and engineers. Growing from this cooperation, the contents of this book draw from the MAD Blockchain 2020 and MAD Artificial Intelligence 2020 conferences that continued to gather industry and academic professionals from a broad spectrum of creative and scientific fields of study. The work that they presented explored blockchain and artificial intelligence in the context of digital games.

This publication is divided into two parts that cover two aforementioned major fields of innovation. Initially, several blockchain items are presented, followed by a showcase of AI use cases and research. Part 1 highlights unique aspects of blockchain technology and discusses case studies that underline how blockchain technology can be used in games, education and society in general. Joshua Ellul and Gordon J. Pace combine the capabilities of blockchain and the classical design of Nomic game to generate insights about self-regulating organizations of players. Daniel Kraft tackles the challenge of blockchain scalability by decoupling the game state from the core network. A collaboration between several Austrian universities and companies, represented by Thomas Wernbacher and his colleagues, presents the results of developing and testing a blockchain-based application that gamifies reporting on waste heat in the city. Andre Thomas and Alexander Pfeiffer present the prototype of their educational game on blockchain that enhances learning experience in the history of arts. J. Tuomas Harviainen, Alesha Serada and Tanja Sihvonen collaborate to present the multidimensional picture of blockchain technologies in commercial games and virtual worlds. Zooming into particular design challenges of such games, Alesha Serada evaluates blockchain technologies as the basis for fairness in play. Finally, Tanja Sihvonen, Merja Koskela and Liisa Kääntä describe the results of linguistic analysis performed on the topics of blockchain and artificial intelligence in Finnish journalistic texts.

Artificial intelligence has gained a lot of prominence in the last decade, and it would be very unlikely to meet someone who has not heard of it. The scalability with which artificial intelligence has grown is attributed to decades of previous work pushed by machine learning pioneers such as Geoffery Hinton, Yoshua Bengio and Yann Lecun, whose work has been actualized through hardware improvements. The state of the area today has led to all major industries, including the digital games sector investing heavily into the use of artificial intelligence applied to different facets of the industry, from automating game development processes to developing autonomous agents that are able to challenge human players better. This part of this anthology presents a number of technical works, such as the work presented by Siracuse et al. and Sammut Bonnici et al., that illustrate the use of generative adversarial networks and reinforcement learning, respectively. Neundlinger et al. also present their interesting work that combines virtual reality and artificial intelligence to understand the limitations of conversational artificial intelligence and provide insights on how to design immersive VR training for social skills. This part also includes work that delves into ethical issues, an area of research increasingly becoming more popular as the use of artificial intelligence as well as the justification of how it comes to certain conclusions is being questioned further. This part includes the work presented by Adnan that explores how AI can be encompassed through a legal framework to protect citizens from misuse and how the ethics of care can be applied to AI. Another viewpoint on ethics and AI is also presented by Giulia Taurino that attempts to present an approach to shape AI practice and understand how information filtering systems transform media cultures. Finally, Alexander K. Seewald presents work exploring the future of AI and how further development in this area will take us toward artificial general intelligence.

This volume is a collection of recent research that was made possible through the collaboration of multiple researchers and academics in the industry. The editors would like to take the opportunity to thank all the participants that attended the conference and contributed to the book proceedings. Special gratitude goes to Nicholas Jushchyshyn and Michael Wagner from Drexel University, Philadelphia, Scot Osterweil from the MIT Education Arcade, Natalie Denk from Danube University, for their ongoing support with the conferences and the anthology.

Msida, Malta Mark Bugeja

Contents

AI

Blockchain

SoliNomic: A Self-modifying Smart Contract Game Exploring Reflexivity in Law

Joshua Ellul and Gordon J. Pace

Abstract Smart Contracts built on Blockchain systems have brought about the ability to deploy code that is tamperproof, verifiable and guaranteed to do exactly what it is written to do over a network of decentralised systems. This enables for decentralised (and centralised) organisations to define rules for particular operations, or even for the entire organisations' decision-making processes. The rules, decision making processes and ability to alter both the rules and processes are encoded within smart contracts. Nomic was a game of self-modifying rules originally meant to explore self-reference and amendment from a regulatory perspective. In this paper we explore the encoding of such a rule system formally as self-modifying yet otherwise tamper-proof code in smart contracts.

Keywords Blockchain · Smart contracts · Nomic · Normative reasoning · Blockchain game

1 Introduction

The notion of self-reference has long been recognised as a source of paradox, and can be traced back at least to Epimenides' paradox[1] if not even earlier with various religious concepts being rife with self-reference, particularly in creation myths. Philosophers, logicians, and later on, mathematicians have formally explored and studied such self-referential systems recognising that the complexity induced by such a notion is best avoided in a formal system if at all possible. However, many

[1]Epimenides claimed that "All Cretans are liars," despite being himself a Cretan.

J. Ellul (✉) · G. J. Pace
Centre for DLT, University of Malta, Msida, Malta
e-mail: joshua.ellul@um.edu.mt

G. J. Pace
e-mail: gordon.pace@um.edu.mt

Department of Computer Science, University of Malta, Msida, Malta

A. Dingli et al. (eds.), *Disruptive Technologies in Media, Arts and Design*, Lecture Notes in Networks and Systems 382, https://doi.org/10.1007/978-3-030-93780-5_1

domains do not have the luxury of choosing whether or not to allow self-referential statements. It suffices to note that two of the greatest mathematical results of the twentieth century, those of Gödel's incompleteness theorem [3] and Turing's results on computational limitations [12] arise directly due to the inescapability of self-reference in mathematics and computation.

One other domain in which self-reference is unavoidable is that of legislation and regulation. Literature on the philosophy of law and normative logics is rife with instances of paradoxes arising from such self-reference. In particular, regulatory documents which regulate the way they may be modified cannot escape self reference. Not all rule systems require this feature—it suffices to look at the rules of the vast majority of games which regulate how players interact but do not provide means through which the rules can be changed as part of the game itself. A notable exception to this rule is that of Nomic [4], a game developed by Peter Dain Suber in 1982 precisely with the intent of exploring self-amendment, thus embracing rule modification at the very core of the game. One may argue that since then, a number of games have explored player interaction with the rules, from the ubiquitous 'The text on the cards trumps the rules' from many a card game to legacy games which allow players to modify the rules as the game progresses, even if this is largely limited to having first class rules which are immutable, and second class rules with which players may interact. Where rule self-amendment starts and where it stops is nebulous. Consider a re-ruling of snakes and ladders in which (i) players are required to move forward one step during their turn; and (ii) players roll a six-sided die at the beginning of their turn, thereby amending the previous rule to say "players are required to move forward n steps during their turn", where n is the number shown on the die. Clearly, the ruleset is not universally amendable (the second rule cannot be changed), but it clearly has an element of self-amendment (at the ruleset level) albeit lacking player agency.

Perhaps Nomic can be seen as the maximal element in the ordered set of games with self-amending rules, but even this is unclear. Nomic can also be seen to be the all-encompassing game, containing within it the possibility to play any other game—perhaps other games can be seen to be 'bonus games' that are possible within the main game. Nomic allows for rules to be fixed in stone: it suffices that the players agree to enact a zeroth rule saying 'Notwithstanding what any other rule of this game may say, this rule cannot be modified or removed'. Legal discussions aside as to the enforceability of such a rule, perhaps an even freer game is a variant of Nomic in which no rule may be set in stone. Then again, this maxim is itself an immutable (meta?) rule of this variant of Nomic contradicting itself. It is clear that deciding the hierarchy of rule amendability in games is far from easy to define and reason about.

Another element which is largely overlooked in the discussion of Nomic (and other self-amending rulesets) is that such a ruleset resides in an ecosystem of social, economic and technological norms. Players of a game frequently reside to social norms, for instance when they realise that they have misinterpreted a rule and agree to proceed with the game abiding by the misinterpreted rule or when players decide to adopt house-rules which they may deem to be more interesting or fun. Ambiguity in rules typically require players to agree on an interpretation—and yet nothing in the rules of Nomic states how such an agreement must be reached[2] and the social contract between players has to kick in. What stops players from agreeing to ignore a Nomic rule 'for the time being'? Such regulatory frameworks sitting above normative systems similarly exists with legal systems—no revolution ever considered whether the changes it is striving for abide by the current legal system.

In this paper we explore an implementation of Nomic as a set of smart contracts written in Solidity (hence the name). Smart contracts [11, 13], residing on a blockchain [8], or similar distributed ledger technology (DLT) have been hailed as the computational embodiment of legal contracts. The implementation allows for an unambiguous operationalisation of the game rules and the process of self-modification. However, its value lies even more as a vehicle for the exploration of issues and underlying assumptions inherent in self-amending systems.

The paper is organised as follows, in the next section we'll briefly describe what blockchain and smart contracts are followed by a discussion highlighting aspects that can be seen as games. In Sect. 3, we will provide implementation details of SoliNomic, and will follow with a discussion in Sect. 4. Finally, Sect. 5 will provide concluding thoughts.

2 Blockchain, Smart Contracts, Rules, and Games

Since Bitcoin was proposed in 2008 [8], blockchain and related DLTs have been proclaimed as game-changing (no pun intended) technology—allowing for centralised points of trust to be removed. This is achieved by replicating data storage and repeating verification and auditing processes across all nodes in the decentralised network (at least in the traditional platforms). To overcome the double-spend problem (which allows for someone to spend their same last unit of currency at two different points in the network at the same time), proof-of-work was proposed—which provides for a means to: (i) select a node in a random-like manner to be the next node to propose a block of transactions to add to the canonical ledger (so that it would be impossible for hackers to determine which node to attack at any point in time); (ii) allow for the time when blocks are to be added to the ledger to be regulated; (iii) allow for nodes to easily check that the block is valid; and (iv) eventually agree on which blocks to accept as truth in the case of different blocks being added at the same time. This is what a blockchain achieves, a means of creating a network of untrusted computers which can maintain a ledger (or append only database) without a single trusted node—unlike distributed systems that had been proposed prior.

Bitcoin provided for the implementation of a ledger tracking cryptocurrency transactions and ownership, yet Blockchain and DLTs can be used for more than that of cryptocurrencies. Ethereum [13] proposed exactly this—a general purpose blockchain that would allow for different applications, typically written in Solidity,

[2] Not even adopting a rule regulating how to resolve ambiguity solves the problem if an ambiguity is found in this very same rule.

to execute on top of it. These applications that execute on blockchains are often termed smart contracts. Mainly because one of their main features are that the code is transparent and available for anyone to view and no one can manipulate code that has already been deployed—so whatever is written in the code is guaranteed to be executed and therefore can be seen as the execution of promises encoded within. Later in Sect. 3 we'll demonstrate snippets of our implementations in Solidity—ample material can be found online to get started with Solidity and was left due to space restrictions.

At the blockchain level, the different nodes can be seen to be working together to maintain the ledger and ensure that only new valid blocks of transactions can be added, and old valid blocks can neither be removed nor edited—some might say following strict rules. Really though, not all nodes are working together in a benevolent manner and neither are nodes necessarily following strict rules as per the specified blockchain protocol. Actually, a blockchain could be viewed as a game, made up of: (i) benevolent players that are correctly following the protocol; (ii) those that may have bugs in their code which could work against the aim of 'correctly' upkeeping the ledger; and (iii) malicious players aiming to invalidate or alter historical transactions. If either (ii) and/or (iii) acquire a coordinated majority[3] of computational power in the network then it would be possible for an attack to be undertaken in which case the benevolent players would have lost in ensuring the ledger's validity. On smaller sized networks it is more feasible to implement such an attack. On a large network like Bitcoin it becomes infeasible.

Whilst, at the blockchain level, rules to maintain the ledger cannot be enforced, by ensuring enough benevolent players (with enough computational power) are in the game, then the ledger's validity will be sustained. With systems like Ethereum, that allow for different application logic to execute on top of the blockchain (instead of just tracking cryptocurrency transactions), any rules can be encoded. At the abstract level of smart contracts, a valid ledger is assumed (and should be taken care of by the blockchain). As long as the blockchain keeps doing what it should be doing (maintaining the valid ledger), then the smart contracts will continue to execute the code that is encoded inside of them. So, the hardcoded rules encoded within smart contracts cannot change, unless the smart contracts originally factored in rule changes—in which case the original rules would still not have changed, since the original rules include rule changes. Within smart contracts, any game could be encoded (subject to expressibility and/or computationability limitations), allowing for different types of players and rules. The rest of this paper will focus on the implementation of SoliNomic, nomic coded in Solidity allowing for different games to be encoded and evolve within the Smart Contract level.

[3] Though Saad et al. [10] cites that less than 51% may be required.

3 SoliNomic

The rules of a game with a static ruleset can be operationalised to provide functionality regulated via smart contracts. The rules would thus be enforced by the logic written on the smart contract, ensuring that players cannot cheat by unilaterally changing the state of the game or playing outside the rules. The immutability of smart contracts ensures that the rules are there to stay, and cannot be tweaked halfway through a game.

3.1 Representing the SoliNomic Ruleset

Representing Nomic in terms of a smart contract thus leaves one in a quandary since at the very heart of Nomic is the axiom that the ruleset will change. One possibility is to represent the ruleset as data (e.g. an expression in a domain specific language rich enough to express any potential behaviour described in the ruleset as it evolves) in which it can (i) be interpreted by the SoliNomic smart contract to execute the logic of the rules, but also (ii) modified by the logic itself as all data stored in a smart contract can. This detaches the isomorphism between the ruleset and the smart contract, since the correspondence would now be between the ruleset and the state of the smart contract, with the smart contract logic acting just as an interpreter of the data. Our intention was to preserve the link between the ruleset and the smart contract, with the latter being a direct operationalised version of the former. Since, however, one cannot update a smart contract's code (bytecode) after it is deployed on the blockchain, we use a standard design pattern for updateable logic—that of allowing the control of a smart contract to migrate to a new one which may have different logic. If such changes are only accepted if invoked by the current ruleset smart contract itself, we are guaranteed that any changes would be according to the current rules of SoliNomic and cannot be changed outside their restrictions.

3.2 The Game Versus the Ruleset

The identity of a non-reflexive game can be associated with its ruleset. The game of chess corresponds to the rules of chess. But this correspondence is lost on reflexive rulesets. If we were to use such a correspondence, a game of Nomic would morph into a new game (since the ruleset has changed) the moment a move is made. But the players would still see themselves playing as the same game instance. We have decided to make this distinction between a game instance and the current ruleset explicit in the implementation of SoliNomic. An overview of the architecture is depicted in Fig. 1. At its heart, a SoliNomic game instance is a smart contract that keeps a reference to the current ruleset, and provides functionality to allow for the ruleset to be updated, as shown below:

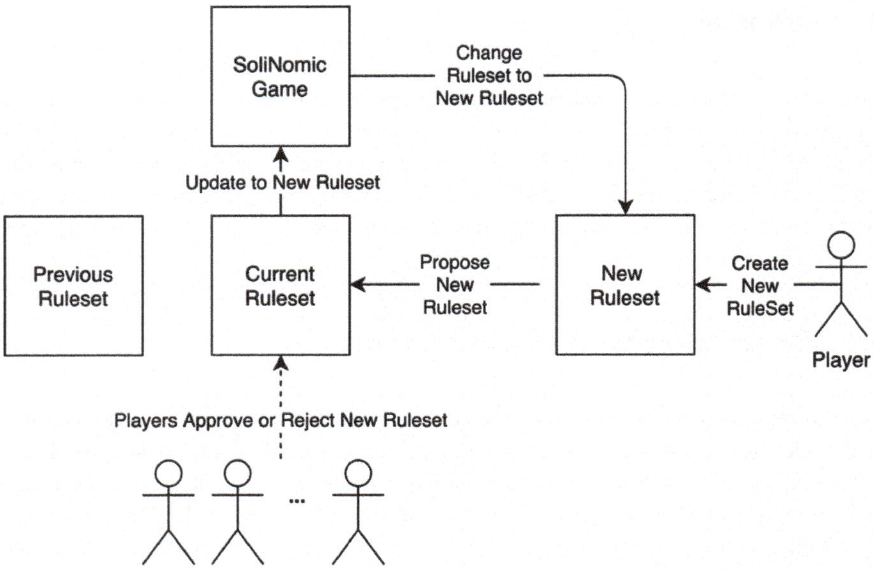

Fig. 1 SoliNomic overview

```
contract SoliNomicGame {
    // Address of the ruleset
  address ruleset public;

  function updateRuleset(address _ruleset) public {
      require (msg.sender == ruleset);
        ruleset = _ruleset;
  }
}
```

Note that the game smart contract does not embody any game logic, which is reserved to the ruleset smart contract. The only logic in the game is that of allowing updates to the ruleset emanating from the current ruleset smart contract itself. The ruleset smart contract would, in turn, embody the operationalisation of the logic, and route any changes to the ruleset to the game smart contract as follows.

```
contract SoliNomicRuleset0 {
  // Address to the game to which this ruleset belongs
  address game public;
  ...

  // Players can propose a rule change (represented by
      new ruleset smart contract)
  function propose(address _proposedRuleset) public {
    ...
  }

```

```
  ...

  // Adopt a proposed change if the majority agree
  function approveAmendment() public {
    // Ensure that the majority have voted in favour
       of the proposed
    require (2*votesInFavour > playerCount);
    // Abdicate in favour of the new ruleset
    game.updateRuleset(proposedRuleset);
    selfdestruct();
  }
}
```

This architecture already raises various implementation options, which run parallel to interpretations in the real world Nomic. Consider the list of players in a game. Given that the ruleset can be changed to enroll new players or drop old ones, should the list of players be stored in the state of the ruleset, or is it an inherent part of the game, and should thus reside in the game smart contract with logic to allow the ruleset to update the list of players? From an implementation perspective, both are possible, with the former resulting in simpler logic (since there is no need of player manipulation to be encoded at the game smart contract level) whilst the latter can be argued to be closer to what the real world game. However, since the game rules proposed for Nomic simply refer to 'the players' with no rules to cater for adding or removing players, neither option is the absolutely faithful one. In SoliNomic, one can experiment with either option to investigate the difference in feel of the two options.

3.3 *Rules Versus the Ruleset*

Amendments in Nomic are meant to be a single rule, even if this arises only indirectly from 111, which states that 'if [a rule-change] arguably consists of two or more rule-changes compounded [...] then the other players may suggest amendments or argue against the proposal before the vote'. Although (arguably) still giving leeway to multiple amendments (if the other players choose not to amend or argue against it), the intention is clearly that single rule changes. In contrast, the code shown above allows for 'global' amendments—proposed amendments to the rulebook as a whole. We have also explored the option of factoring the ruleset into individual functional elements which can only be amended, added or removed one at a time.

Each rule clause can be encoded as a simple smart contract with a standard means of invoking it:

```
contract Clause17 {
  ...
  function execute(...) public {
    ...
  }
}
```

The individual rules are then collected together in a ruleset contract with a mapping from rule name[4] to the rule contract and allows for invoking of individual clauses, and for amendment of individual rules from the game contract:

```
contract SoliNomicRuleset0 {
  ...
  // The rules making up the  ruleset
    mapping (bytes32 => address) rules public;
  function invoke(bytes32 _rulename) public {
      rules[_rulename].execute(...);
  }

    function amend(bytes32 _rulename, address
       _rulelogic) public {
    require (msg.sender == game);
    rules[_rulename] = _rulelogic;
  }
}
```

The game contract remains almost unchanged from the one we saw earlier, except that amendments are now proposed on a particular named clause:

```
contract SoliNomicGame {
  // Address of the ruleset
  address ruleset public;

  function updateRule(bytes32 _rulename, address
     _rulelogic) public {
    require (msg.sender == ruleset);
    ruleset.amend(_rulename, _rulelogic);
  }
  ...
}
```

In this manner, we can achieve rule-centric amendments being proposed by players. Having shown how such a rule-level amendment process can be adopted and implemented, it is worth highlighting that this clause-level amendment could have easily been implemented as an amendment to the ruleset smart contract using the previous approach, with proposals being made and voted upon at the clause level i.e. leaving the rest of the ruleset unchanged. The flexibility of SoliNomic (and Nomic) ensures that the two approaches can result from one another in the course of a single game.

---

[4] For efficiency reasons, we use the hash of the name of the clause, but for all intents and purposes this can be seen as a mapping from clause name to clause functionality.

4 Discussion

Regulating interaction in a decentralized manner: Nomic does not adopt a centralised authority regulating the players with respect to the game rules, but instead depends on the players to regulate their own behaviour. SoliNomic goes one step further in that it provides an automated way in which the rules are enforced, thus ensuring that the players do not diverge from them. However, the mechanism inherent in the smart contracts running on a DLT ensure that there is no central authority controlling them. Although one finds other automated Nomic rule enforcement engines in literature e.g. see [1, 9] and Nomyx[5] these were all implemented on a central server, meaning that the administrator of the server can disrupt or modify the rules.

Games as decentralised (autonomous) organisations: The issue of decentralisation highlighted an aspect of many games which provide regulated interaction between players without giving the power of enforcement to a subset of the participants. The rules are adhered to in a decentralized manner, using social norms to discourage players from cheating, and decentralised enforcement with players observing each other to ensure that everyone is playing by the rules. By adding a decentralised engine to enforce conformance with the rules, as we have done in SoliNomic, we have essentially created a decentralised autonomous organisation (DAO) [2], in which the behaviour is encoded by unambiguous executable rules, yet providing a degree of control by the organization members without being influenced by a central authority.

Tyranny of the majority: Some may argue that the automated enforcement provided by SoliNomic ensures that the game cannot diverge from the rules themselves, along the lines of Lessig's code is law maxim [6]. For instance, one cannot revert back time and undo changes in an arbitrary manner. However, nothing stops the majority of the players from deciding to stop using the original SoliNomic game smart contract and create a new one redistributing points or taking off from a point in the past. This is analogous to how the Ethereum community (controversially) chose to undo a successful hack of the original DAO fund raising smart contract by collectively choosing to resume the Ethereum blockchain from a few blocks back [7]. The existence of a social layer above the automated system leaves space for such divergences, and it can only be the social contract between the members of the organisation that discourages them taking such action.

Pushing the boundary of what is mutable and what is immutable: As we have seen in the previous section, the implementation of SoliNomic brought out more clearly underlying assumptions about mutability and immutability of rules. The choice of having any logic starting in the game smart contract which is immutable ensures that the rules embodied in that logic cannot be modified. However, this can be pushed further by adopting a hierarchy of smart contracts, allowing more mutability the lower down one moves in the hierarchy. We have seen this with named clauses, which allow controlled mutability of the ruleset, and this notion is already present

[5] https://github.com/nomyx/Nomyx.

in the original rules of the game of Nomic which consist of mutable and immutable rules (Nomic, not blockchain terms) and a process of transmutation to allow for rules to migrate from one form to another.

Mutability of state versus mutability of rules: One interesting feature of games which is highlighted by the implementation of SoliNomic is that games offer a degree of mutability of game state (e.g. the position of the pieces in chess), which is distinct from mutability of rules which only a few games permit. However, the former can be modelled as a limited form of the latter, e.g. adapting the chess rules to include information as to where the pieces lie and whose turn it is, and then interpreting the moving of a piece as an amendment to the rules. The dividing wall between rules and game state has been explored in a number of game genres, but the degree to which it is explicit in SoliNomic enables exploration of this space more clearly.

Bugs and features: SoliNomic takes the approach of code is law—what the code does is what the rules say. Note that this is distinct from saying that what the code is intended to do is what the rules say. The notion of bugs in the code does not exist, since the code is by this definition canonical. Features of the game arising from bugs typically cut the game short, allowing the perceptive player to use the behaviour to their advantage. However, occasionally it may lead to interesting unplanned scenarios which diverge from the players' original intent.

Dealing with ambiguity: The fact that Nomic rules are written in a natural language means that this may result in ambiguity which the players would have to resolve, so much so that the original rules explicitly refer to such a possibility. In contrast, an automated implementation means that such ambiguities would have to be resolved when implementing as an operationalised ruleset. We have already discussed player management. The rules simply refer to the players without providing any mechanism to manage who is playing. An inherent assumption may be that the players remain the same throughout the game, but some players may allow for players to join or leave the game as it evolves. Others may choose a midway interpretation in that the players are fixed but one may adopt new rules to allow for onboarding and jettisoning of players. Another source of ambiguity in the English ruleset lies in the proposal and voting mechanism. While the rules are specific in saying that 'Turns may not be skipped or passed', it is unclear how a group of players are to proceed if a player refuses to make a proposal when it is their turn. Similarly, it is underspecified whether all players have to vote, whether the vote is secret or not, whether players vote simultaneously or in sequential order, etc. Operationalising the rules means that one has to make an explicit decision for each of these choices. If different options are to be kept, then there still has to be the explicit choice handled by the logic (e.g. the current player takes the decision of which interpretation to adopt).

Declarative versus operational descriptions: Although the vast majority of smart contract languages are operational i.e. they allow one to express how to achieve a goal, one can find a number of languages which attempt to take a more declarative approach i.e. allowing one to express what is to be achieved as opposed to how to achieve it. Legal texts and game rulebooks are typically declarative, and this leads to a gap between the implementation of SoliNomic and the original ruleset. The only implementation of a declarative approach to Nomic we are aware of is Bananomic [1] which, however, limits the rules to talk about a concrete scenario (rules regulating monkeys which are throwing bananas at each other in a tree) in order to ensure that the declarative rules can be interpreted and executed.

Other attempts at automating Nomic: In Nomic, as is the case with most games, rules are expressed in a declarative, not operational manner. The discussion as to the relationship between legal and smart contracts has long been discussed [5]. Normative rules specify the ideal course of affairs ('One ought to submit their tax return by the end of June of the following year'), but recognise that such ideal behaviour is not guaranteed ('A fine of €10 per day shall be incurred for late submissions of one's tax return'). This is distinct from smart contracts which typically operationalise behaviour to ensure compliance with the rules. It is worth noting that in the case of game rules, despite them being expressed as declarative normative rules, only compliant behaviour can typically be observed: if the rules say that 'A player may play no more than three cards per turn', players do not have the option to violate the rule (in the same way that a person may choose not to submit their tax return by the due date).

Our approach in SoliNomic follows that of various automated versions of Nomic such as PerlNomic [9] take an operational view, equating proposals with code to be executed. In contrast, BanaNomic [1] took a declarative approach, but suffers from a need to understand various actions (e.g. 'play card', 'propose amendment') and states (e.g. 'no player has more than 10 points', 'no amendment has yet been accepted') and their relationship. The operational approach provides a shortcut in that the code refers to and modifies the state of the game directly.

Without the use of smart contracts, previous implementations suffered from the problem of centralised governance of hardware. The owner of the server may choose to change data without going through the rules of the game. Consider PerlNomic, in which the rules took the form of Perl scripts accessible from a server. The scripts provided limited means through which the players were allowed to change the scripts themselves. However, the person owning the server could easily edit these scripts directly without following these rules. Smart contracts ensure that this is not possible.

5 Conclusions

In this paper we presented a decentralised version of nomic, SoliNomic, implemented in Solidity. To the best of our knowledge this is the first decentralised version—which does not require players to regulate or keep in check the actions of others, and neither allows for any centralized system operator or hardware maintainer to manipulate the state and/or game rules. Further work should be undertaken to investigate other design options in regards to: (i) which logic and/or rules should be well-defined in the smart contract code and which should remain as unstructured logic left up to the players to define; and (ii) different design patterns for updating rulesets, rules and players.

References

1. Camilleri, J.J., Pace, G.J., Rosner, M.: Controlled natural language in a game for legal assistance. In: International Workshop on Controlled Natural Language, pp. 137–153. Springer (2010)
2. Chohan, U.W.: The decentralized autonomous organization and governance issues. Available at SSRN 3082055 (2017)
3. Gödel, K.: On Formally Undecidable Propositions of Principia Mathematica and Related Systems. Courier Corporation (1992)
4. Hofstadter, D.R.: Nomic: a self-modifying game based on reflexivity in law. In: Metamagical Themas: Questing for the Essence of Mind and Pattern, pp. 70–86. Bantam, New York (1985)
5. Ladleif, J., Weske, M.: A unifying model of legal smart contracts. In: International Conference on Conceptual Modeling, pp. 323–337. Springer (2019)
6. Lessig, L.: Code: and other laws of cyberspace. ReadHowYouWant.com (2009)
7. Mehar, M.I., Shier, C.L., Giambattista, A., Gong, E., Fletcher, G., Sanayhie, R., Kim, H.M., Laskowski, M.: Understanding a revolutionary and flawed grand experiment in blockchain: the DAO attack. J. Cases Inf. Technol. (JCIT) **21**(1), 19–32 (2019)
8. Nakamoto, S.: Bitcoin: a peer-to-peer electronic cash system (2008). https://bitcoin.org/bitcoin.pdf
9. Phair, M.E., Bliss, A.: PerlNomic: rule making and enforcement in digital shared spaces. Online Deliberation (2005)
10. Saad, M., Spaulding, J., Njilla, L., Kamhoua, C., Shetty, S., Nyang, D., Mohaisen, A.: Exploring the attack surface of blockchain: a systematic overview. arXiv preprint arXiv:1904.03487 (2019)
11. Szabo, N.: Smart contracts: building blocks for digital markets. EXTROPY J. Transhuman. Thought (16) **18**(2) (1996)
12. Turing, A.M.: On computable numbers, with an application to the entscheidungsproblem. Proc. Lond. Math. Soc. **2**(1), 230–265 (1937)
13. Wood, G., et al.: Ethereum: a secure decentralised generalised transaction ledger. Ethereum Proj. Yellow Pap. **151**(2014), 1–32 (2014)

HotCity—A Gamified Token System for Reporting Waste Heat Sources

Thomas Wernbacher, Alexander Pfeiffer, Ernst Gebetsroither-Geringer, Marko Goels, Jakob Worster, Ernst Meißner, Angelika Graf, Romana Stollnberger, Roman Geyer, Ralf-Roman Schmidt, and Alesha Serada

Abstract Urban planning needs to discover and incorporate new energy sources to meet climate protection targets in the future. Waste heat from industrial and urban infrastructure has proven to be a viable solution, but its proper identification can be challenging, especially for smaller and unconventional sources. Our project relies

---

T. Wernbacher (✉) · A. Pfeiffer
Donau-Universität Krems, Dr. Karl-Dorrek-Straße 30, 3500 Krems, Austria
e-mail: thomas.wernbacher@donau-uni.ac.at

A. Pfeiffer
e-mail: alexander.pfeiffer@donau-uni.ac.at

E. Gebetsroither-Geringer · R. Stollnberger · R. Geyer · R.-R. Schmidt
AIT Austrian Institute of Technology GmbH, Giefinggasse 4, 1210 Vienna, Austria
e-mail: ernst.gebetsroither@ait.ac.at

R. Stollnberger
e-mail: romana.stollnberger@ait.ac.at

R. Geyer
e-mail: roman.geyer@ait.c.at

R.-R. Schmidt
e-mail: ralf-roman.schmidt@ait.ac.at

M. Goels · J. Worster
Digitalsunray Media, Gonzagagasse 11/ 25, 1010 Vienna, Austria
e-mail: marko.goels@digitalsunray.com

J. Worster
e-mail: jakob.worster@digitalsunray.com

E. Meißner · A. Graf
Grazer Energieagentur Ges.m.b.H., Kaiserfeldgasse 13, 8010 Graz, Austria
e-mail: meissner@grazer-ea.at

A. Graf
e-mail: graf@grazer-ea.at

A. Serada
University of Vaasa, Wolffintie 34, FI-65200 Vaasa, Finland
e-mail: aserada@uwasa.fi

A. Dingli et al. (eds.), *Disruptive Technologies in Media, Arts and Design*, Lecture Notes in Networks and Systems 382, https://doi.org/10.1007/978-3-030-93780-5_2

on the principles of gamification enhanced by a blockchain based token system and crowdsourcing as methods to collect and utilise spatial data such as the location and the size of previously unused heat sources. The mobile platform-neutral HotCity App en ables users to collectively patrol the city in search of waste heat sources and to gain tokens that can be exchanged for rewards. The blockchain platform Ardor was used for cheat proofing and to enable transparency for the reward system. The field test conducted in winter 2020/2021 showed high usability scores as well as high acceptance ratings of our approach opening up new use case scenarios in the context of spatial energy planning.

Keywords Blockchain · Blockchain game · Gamification · Energy saving · Crowdsourcing

1 Introduction

Availability of reliable data is one of the most important elements for fact-based decisions. Urban planning and spatial energy planning often suffers from a lack of availability of good, validated and up-to-date data sets. Furthermore, integrated spatial and energy planning needs to incorporate new spatially distributed energy sources and understand how these sources can be used in the future to meet climate protection targets. These new energy sources can be, for example, waste heat from industrial food production, local industrial/commercial enterprises, data centers, or urban infrastructure such as tunnels and metro stations. The utilization of such waste heat sources in heating networks has been demonstrated several times, however, their proper identification in an urban environment can be challenging, especially for smaller and unconventional sources [17].

Gamification as an innovative way to collect the needed data was investigated within a national funded research project “HotCity”. Gamification builds on the use of game mechanics in contexts that are, by nature, unrelated to the game [7]. Within the project the “HotCity App” was developed enabling users to spatially report and evaluate different sources of waste heat. The gamification of data collection was also intended to raise awareness of waste heat and energy use on the one hand, and to facilitate the collection of data from small energy sources on the other. For the first time, the game framework is secured using a blockchain and mapped by means of a token system. The HotCity App was tested in the Austrian cities Vienna and Graz as a proof of concept to analyse if and how the gamification approach can deliver valid results.

The HotCity App makes it possible to identify and georeference also smaller sources of waste heat in order to use the available energy. Two test runs were each conducted in Graz and Vienna, which provided helpful feedback from the testers regarding promising features as well as showed barriers reducing the success of this data collection approach.

The paper will elaborate the game development, discuss the gamification approach and the lessons learned during the proof of concept project.

2 Gamifying the Challenge

The energy system of the future will most likely consist of many different decentralised units (e.g. rooftop PV systems, individual heat pumps, etc.). For the development of districts with high energy efficiency and increased use of locally available and sustainable energy sources, a detailed spatial identification of possible energy potentials is necessary in order to plan cost-efficiently and future-proof. In particular, waste heat from industry (foundries, food production …) and commerce (data centres, supermarkets …) as well as urban infrastructure (tunnels, metro stations) can make an important contribution to heating and hot water in plus-energy neighbourhoods. While "low-hanging-fruits" such as waste heat from large industrial plants are already used on a large scale, the identification of smaller sources is associated with various difficulties [17]. Many larger cities already have a dataset, e.g. Open Government Data, but it is usually not up-to-date enough, not (spatially) detailed enough and often does not contain the necessary data. The often chosen top-down methodology for determining waste heat potential, i.e. via a survey of the largest pollutant emitters (e.g. [3]), does not cover smaller sources, which thus do not appear in these databases.

Gamification, on the other hand, offers the possibility of generating targeted incentive systems for data collection (crowdsourcing/crowd collecting). Games like "Pokemon Go" have shown the unimagined dynamics that can be generated. As part of the HotCity project funded by the Federal Ministry for Climate Action, Environment, Energy, Mobility, Innovation and Technology (BMK), the HotCity App was developed that allows users to spatially record and evaluate new sources of waste heat. The application for the collection of waste heat was chosen as an example for other spatial energy data. The gamification of waste heat data collection was also intended to raise general awareness of the issue and facilitate the collection of data from small energy sources that are not normally considered in top-down approaches.

2.1 Gamification as a Method for Incentivizing the Identification of Waste Heat Sources

The massive distribution and market penetration of digital games (approx. 37 billion dollars in sales in 2020 in the USA alone as stated by [5]) is impressive. For example, 510 million people are currently actively playing games (5.3 million in Austria as stated by [14]), with an average age of 35. Gamification is based on the use of game mechanics in contexts that are, by nature, unrelated to games [7]. The aim of gamification usually is to apply the motivational and feedback techniques that have been tried and tested in games. Games provide clear goals ([10]; e.g. in the form of quests), they reward ([19]; e.g. badges, level-ups), they allow to compete or cooperate with others ([20]; e.g. in the form of rankings, multiplayer elements) and they provide an interactive framework for different experiences and skills [11, 12].

Gamification has already been successfully used in various application areas to promote participation, such as in the context of civic courage [6], citizen participation [18], e-learning [2] and e-government [1]. The application of gamification has also yielded positive results in the mobility sector, e.g. in terms of promoting sustainable forms of mobility [13].

Pfeiffer et al. [15] coined a new definition of Gamification: "Gamification is the use of game mechanics as a further dimension within and around both gaming and non-gaming contexts, in an endeavour to nudge participants to perform certain actions, by adopting a playful attitude". This definition shows that gamification can also take place within games, whenever further elements outside the core mechanics and the core storyline are used to get the players to explore certain content. Furthermore, the definition also shows the relation to the principle of nudging and that gamification can be seen as one of the tools to trigger positive behaviour.

2.2 Blockchain as a Method for Securing the Gamification Framework

Due to its decentralisation, transparency and security, the blockchain is dominating technological and social discourse [4] and is being treated as a disruptive innovator for a wide range of applications: from transaction processing to land registry entries to logistics chains, the intermediary is to be eliminated in the future [9]. Previous crypto technologies such as Bitcoin and the first version of Ethereum rely on the "proof of work" architecture. A small percentage of the cryptocurrency in the network is used to reward "miners", who keep the entire network alive and validate all transactions, for solving randomly generated computing tasks. The computing power required for this increases linearly with the difficulty of the computing tasks and consumes more and more resources, which leads to grounded environmental concerns [8]. This is where new blockchain technologies such as "Ardor" or "NXT" come into play. These solutions are based on the "proof of stake" algorithm and are much more energy-efficient. In this case, it is not about computing power in the form of graphics card performance or CPU-power, but about holding a certain stake in the network-maintenance token, as well as other aspects such as the total time of being active online as a verification node and a certain degree of randomness. Regardless of the particularities of architecture, the peer-to-peer principle, the strong encryption and the permanent and validatable storage of information represent an opportunity for diverse industries—including, of course, the gaming industry [16].

Ardor offers the possibility to generate utility tokens on the child chain Ignis. Approval models can be set up around these created tokens and therefore the set of rules can be specified. Furthermore, there is the possibility to develop smart contracts. As a bonus, there is also a ready-made marketplace that can be used within the scope of the decentralised applications (DApps) generated using the Ardor blockchain and its child chain Ignis.

In HotCity, the blockchain was used to prevent cheating on the one hand and to ensure transparency for the players on the other. It was used to automate the management of redeemed prizes and rewards and to reduce the administrative effort to a minimum. The concept involved incentives (such as coffee vouchers)—mapped on a blockchain token—that could be redeemed at project partners. Double spending or counterfeiting of vouchers can thus be avoided. In addition, the accounting settlement can be automated. For example, similar to a current account system, the partners could send the tokens to our billing account on a monthly basis, thus triggering the transfer of Euro funds from the game operator to the prize provider.

3 The HotCity App

A platform-neutral app design was created for the HotCity App in order to appeal to users from both worlds (iOS and Android). This platform-agnostic design also offers the possibility of drawing users more strongly into the Hot City world, as the branding elements from the style guide are implemented everywhere. For a quick overview, explanatory screens were developed for rapid onboarding, which users are shown at the start of the app and can also view again and again while using the app. The app is based on learned UI patterns (e.g. menu navigation, call-to-actions, map with markers), so that no specific knowledge or explanations are necessary to use the app (Fig. 1).

The home screen of the app or main game area is the overview map. The map is divided into rectangular districts that can be conquered by users and crews. From there, the user has an overview of everything. The user sees the city from a bird's eye view, above it the districts and the heat sources. The heat sources have a status that shows how much information can still be added by users. Users can add new heat sources directly from the home screen via the "Upload photo" function. The photo is first saved and then entered on the Basemap Austria via a map picker. In the next step, the user is asked to add first details to the found heat source.

By earning more heat tokens, users can climb up the ranking in the app's leaderboard and earn new badges. Besides, they can form crew and engage in cooperative play. In this scenario, each user and each crew aims to occupy predefined squares according to the plan in the app and thereby both increase their heat score and earn new badges.

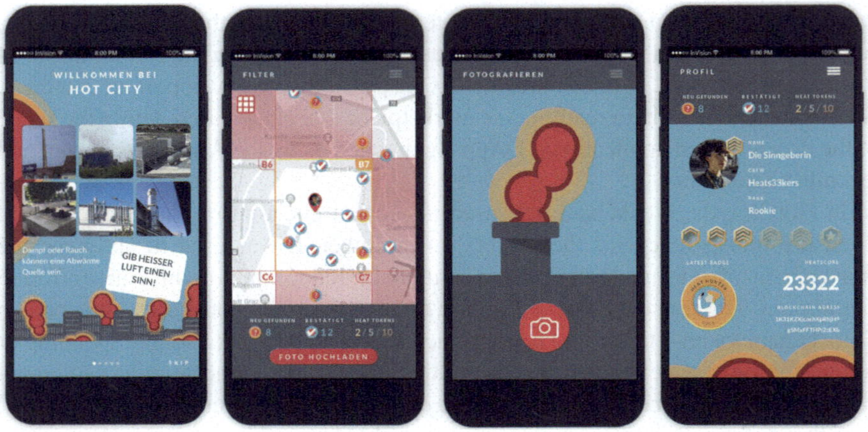

Fig. 1 Screenshots from HotCity App as they would look on the user's device

3.1 The Token Framework

The gamification framework consists of three utility tokens realised within the blockchain Ardor and its child chain Ignis that represent points: bronze, silver and gold tokens. Users receive bronze tokens for simple tasks such as uploading unconfirmed images and later receive points with higher value, when heat sources are confirmed by other payers. These tokens can later be redeemed for incentives, with better incentives distributed for gold tokens than for bronze tokens. As a further game mechanic, tokens can be exchanged for each other, e.g. 10 silver tokens result in 1 golden token. Furthermore, all badges are displayed as tokens.

The actual transfer of the respective tokens in the form of assets takes place entirely on the blockchain. A lightweight contract (cf. smart contract) was developed for this purpose, which, in conjunction with the API, ensures that the assets are transferred correctly. This contract is initially triggered by the backend and contains, for example, information about which "heat spot" has triggered an asset transfer. The lightweight contract then calls the API and checks whether the entries and information are valid and correct and then transfers them. In a freely definable period of time, the API checks whether the asset transfer was successful. If, for example, the transfer of the assets fails due to technical circumstances or insufficient fee, a new process takes place. The Ardor node primarily manages and executes the lightweight contracts for the transfer of the application. As already mentioned, the API acts as a validation unit for this contract and only transfers assets or badges if the correct information is available.

In summary, blockchain technologies in the HotCity project secure those relevant parts of the game where values for the players or project partners are affected. On the one hand, tokens, badges and vouchers earned and, on the other hand, the accounting modalities. All other data, such as the description of the heat sources, are saved in a conventional database, which is of course designed in the best possible way and continuously backed up. Transactions on the blockchain, as accessed from the graphical interface of the main account full-node as well as a triggered badge.

4 The HotCity Field Test

The field test took place in two different phases during the winter 2020/2021. The first phase was performed in November 2020, while the second phase was launched mid of February 2021 for 3 weeks. For a part of the test persons, fees were awarded for 5–8 h playing time. All participants received a briefing in advance in the form of a tutorial on the app and in the form of marked districts that showed a high potential in terms of waste heat sources. The responses were collected in anonymous form with a request for open critical feedback. A total of 836 tokens were sent out for (a) heat sources, divided into bronze, silver and gold and (b) badges (Fig. 2).

After a 2–3 week test phase, an online questionnaire was created and submitted using the "Survey Monkey" software. The item collection contained both open and closed questions with dichotomous and rating scaled gradations. A total of 31 people took part in the test and filled out the online questionnaire. The majority of the test persons (almost 68%) tested the app in Vienna, the rest (32%) in Graz.

4.1 The Game Rating and Winning Strategies

Satisfactory values could be reported with regard to the usability and acceptance of the app. The usability is subjectively high (mean values from 2.03 to 2.62). The simple user interface (M = 2.07) and the assessment of the comprehensibility for other users (M = 2.03) are particularly noteworthy in this regard. The normalised SUS (system usability scale) score is 79.675, which corresponds to an above average value (maximum = 100, minimum = 0).

A comparison between the first and second test phase shows that incremental improvements to the tutorial led to heat sources being found more easily (M = 2.2 in the second test phase). Solely the input of the additional information should still be improved. Game enjoyment was rated as high (M = 1.83 in the second test phase) (Fig. 3).

Going for a walk was named as the most successful or popular game strategy (27%) because of the positive health effects, followed by the targeted search for heat sources in industrial districts (20%) and searching in advance on Google satellite maps (16%) (Fig. 4).

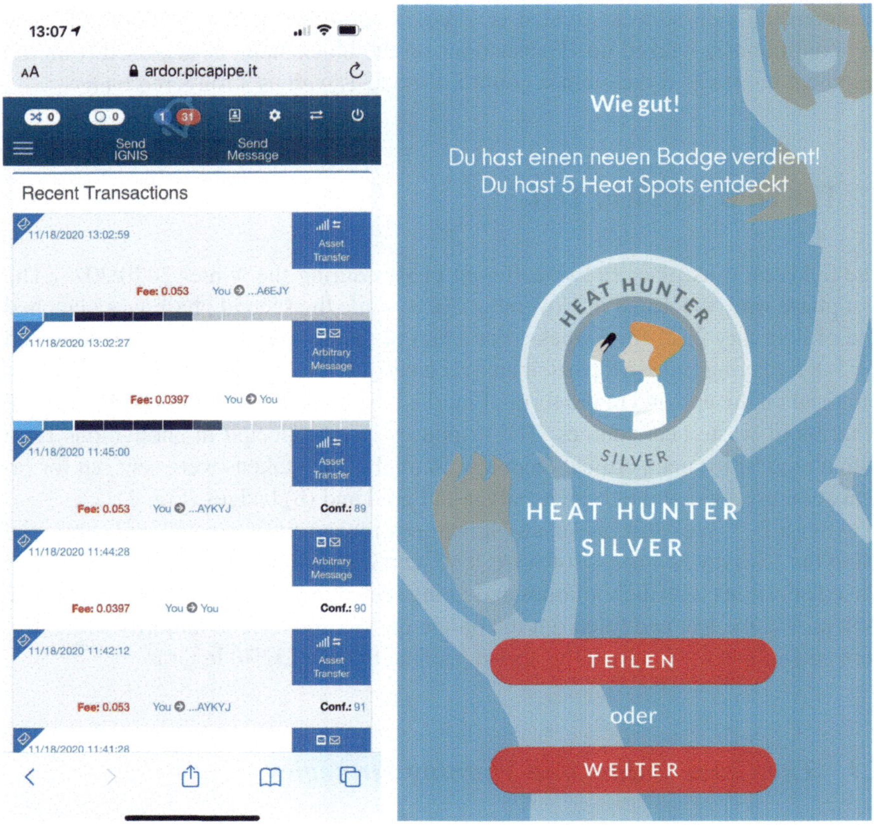

Fig. 2 Blockchain token transactions (on the left) and visualisation of the triggered badge (on the right)

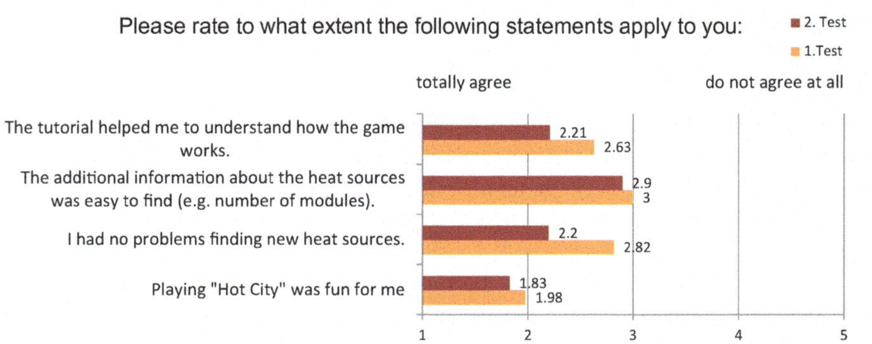

Fig. 3 The game rated by the test players

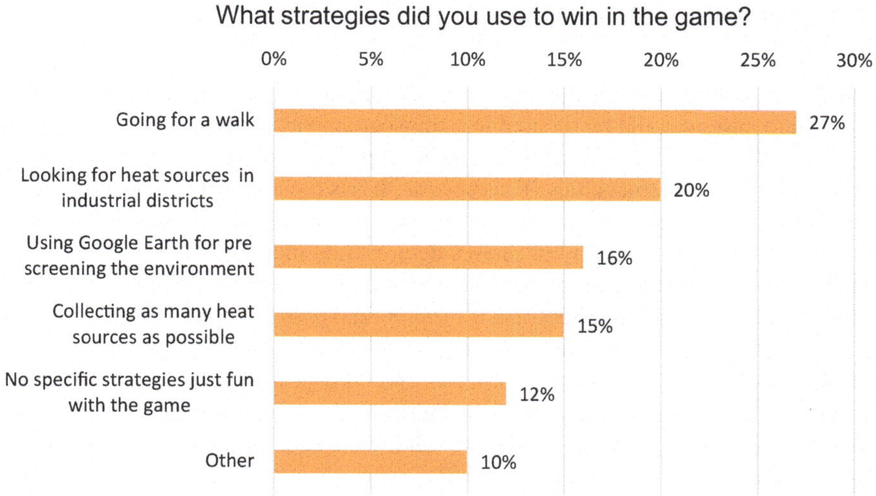

Fig. 4 Winning strategies of the test group

4.2 *Lessons Learned*

The results of the field test show high values for the acceptance of the app. The fun of play and the level of difficulty were also rated as appropriate. The standardised SUS score is in the upper fifth, a high subjective usability is thus evident.

The core functionality of the app—the search for heat sources, which was designed and integrated in accordance with the project goals—is seen as motivating by the users; compared to other game features, the entertainment value is particularly high.

With regard to the qualitative aspects, the design of the app, the game idea and the ease of use were praised by the test persons.

Problems were encountered with regard to GPS positioning, which sometimes did not work immediately and had to be corrected manually (A technical problem here is that automatic GPS positioning reduces the duration of one battery load, thus a compromise has to be chosen, or using manual GPS positioning, which reduces the usability). In addition, as the first phase of the field test in particular showed, it was sometimes difficult for the users to find suitable heat sources, and chimneys from private households were often photographed. Through clearer communication and tutorials, this problem was alleviated in the second test phase.

Suggestions for improvement mainly refer to an improved tutorial (phase I) and an automated location on the map and of the heat sources.

5 Conclusion and Outlook

Gamification can be a promising way to identify urban energy data such as waste heat. The reward system could be secured in a transparent way using the energy-efficient proof of stake blockchain Ardor. The HotCity-App has been successfully developed and tested in two Austrian cities. The two test phases have shown that the app seems to be very promising from a user point of view as the usability score shows. The lessons learned from the two test phases demonstrate that a good briefing (tutorial material) is important to gain useful results within the respective context. Compared to the already existing waste heat data for Vienna and Graz new promising waste heat proposals could be found, which have not been in the data sets so far and for some existing data sets an increased accuracy of the GPS position could be established thanks to the crowdsourcing approach. Interestingly even with the relatively low amount of testers a significant number of valid waste heat proposals could be found. However the tests also helped us to identify potential obstacles especially regarding details attached to heat spot entries which can be used for estimating the amount of waste heat for each waste heat proposal. The results indicate that these are hard to collect with the current app, thus more advanced users (experts) or higher incentives are needed to collect this information. The developed interactive evaluation tool (web-application) seems to be a very good starting point for experts (not necessarily players) in order to filter the most promising waste heat proposals and to estimate their actual potential.

5.1 Recommendations Regarding Gamified Apps for Serious Application Scenarios

Based on the feedback obtained during testing we suggest that similar approaches take into account the following recommendations:

- Provide an optional tutorial video and a half-hour webinar to introduce and explain the app. As user segmentation in the future could include regular users, experts as well as community experts. Such a webinar is an efficient method for onboarding and providing the necessary skills to assess heat sources according to their potential.
- Simplify the registration process by publishing it in the app stores. The current process requires several steps, which cause a time delay for users and thus make "seamless onboarding" difficult. In addition, the release had to be done manually due to restrictive settings on company devices. A release in the app store would prevent such problems.
- Integration of a button to update one's GPS position within the app. This function facilitates correct positioning, and a "double-select" interaction for uploading photos should be introduced so that users can enter or readjust their GPS position correctly.

- Switching the map view to a satellite view. This makes it easier to recognise heat sources, and it is also possible to estimate the height of objects/heat sources using Google Earth 3D, for example.
- Integration of notifications for finding heat sources and invitations to groups. This also regularly encourages users to use the app, but a good balance must be struck so as not to exert a disruptive influence.
- The token logic should be adapted so that the confirmation of heat sources is only possible between different groups/crews. The "report" function for uploaded heat sources should be reframed to create a positive connotation and thus encourage use by users.

5.2 Recommendations Regarding the Use of Blockchain Technology in Gamified Frameworks

With regard to the blockchain, we suggest that future or similar approaches fulfill the following basic requirements:

- It must be a fully decentralised public blockchain.
- Users can, if they want, run a node of the blockchain on one of their devices, such as a laptop or mobile phone.
- We can use approval models to ensure that various conditions are met. One of these conditions is that the tokens cannot be transferred to other unregistered wallets.
- Identity management can potentially be implemented. In our case, this would be a possible connection of decentralized identifiers (DIDs).
- Users do not need cryptocurrencies to pay network fees, we can take care of this as an operator with bundling accounts.
- Individual transactions must be manageable in terms of costs, even if the crypto market fluctuates massively.
- Messages attached to transactions must be encryptable.
- However, it must be possible to generate shared keys. To make the information readable for authorised parties.
- It needs the possibility to programme lightweight smart contracts, in our case it is the exchange of the different token classes as part of the game mechanics.

References

1. Al-Yafi, K., El-Masri, M.: Gamification of e-government services: a discussion of potential transformation. In: AMCIS 2016 Proceedings (2016). https://aisel.aisnet.org/amcis2016/eGov/Presentations/8
2. Barata, G., Gama, S., Jorge, J., Gonçalves, D.: Improving participation and learning with gamification. In: Proceedings of the First International Conference on Gameful Design, Research, and Applications, Gamification '13, pp. 10–17. Association for Computing Machinery, New York, NY (2013). https://doi.org/10.1145/2583008.2583010

3. Brueckner, S., Miró, L., Cabeza, L.F., Pehnt, M., Laevemann, E.: Methods to estimate the industrial waste heat potential of regions—a categorization and literature review. Renew. Sustain. Energy Rev. **38**, 164–171 (2014).https://doi.org/10.1016/j.rser.2014.04.078. https://www.sciencedirect.com/science/article/pii/S1364032114003165
4. Buhl, H., Urbach, N., Schweizer, A.: Blockchain-Technologie als Schlüssel für die Zukunft? Z. gesamte Kreditwes. **70**(Nr. 12), 596 (2017)
5. Clement, J.: Leading gaming markets worldwide in 2020, by gaming revenue. Tech. Rep. Stat. (2021). https://www.statista.com/statistics/308454/gaming-revenue-countries/
6. Coronado Escobar, J.E., Vasquez Urriago, A.R.: Gamification: an effective mechanism to promote civic engagement and generate trust? In: Proceedings of the 8th International Conference on Theory and Practice of Electronic Governance, ICEGOV '14, pp. 514–515. Association for Computing Machinery, New York, NY (2014). https://doi.org/10.1145/2691195.2691307
7. Deterding, S., Dixon, D., Khaled, R., Nacke, L.: From game design elements to gamefulness: defining "gamification". In: Proceedings of the 15th International Academic MindTrek Conference: Envisioning Future Media Environments, MindTrek '11, pp. 9–15. Association for Computing Machinery, New York, NY (2011). https://doi.org/10.1145/2181037.2181040
8. Goodkind, A.L., Jones, B.A., Berrens, R.P.: Cryptodamages: monetary value estimates of the air pollution and human health impacts of cryptocurrency mining. Energy Res. Soc. Sci. **59**, 101281 (2020). https://doi.org/10.1016/j.erss.2019.101281. http://www.sciencedirect.com/science/article/pii/S2214629619302701
9. Hopf, S., Picot, A.: Revolutioniert Blockchain-Technologie das Management von Eigentumsrechten und Transaktionskosten? In: Interdisziplinäre Perspektiven zur Zukunft der Wertschöpfung, pp. 109–119. Springer Fachmedien, Wiesbaden (2018). https://doi.org/10.1007/978-3-658-20265-1_9
10. Hunicke, R., LeBlanc, M., Zubek, R.: MDA: a formal approach to game design and game research. In: Proceedings of the AAAI Workshop on Challenges in Game AI, vol. 4(1), p. 1722 (2004)
11. Ivory, J.D., Kalyanaraman, S.: The effects of technological advancement and violent content in video games on players' feelings of presence, involvement, physiological arousal, and aggression. J. Commun. **57**(3), 532–555 (2007). https://doi.org/10.1111/j.1460-2466.2007.00356.x
12. Jansz, J.: The emotional appeal of violent video games for adolescent males. Commun. Theory **15**(3), 219–241 (2005). https://doi.org/10.1111/j.1468-2885.2005.tb00334.x. https://onlinelibrary.wiley.com/doi/abs/10.1111/j.1468-2885.2005.tb00334.x. eprint: https://onlinelibrary.wiley.com/doi/pdf/10.1111/j.1468-2885.2005.tb00334.x
13. Kazhamiakin, R., Marconi, A., Perillo, M., Pistore, M., Valetto, G., Piras, L., Avesani, F., Perri, N.: Using gamification to incentivize sustainable urban mobility. In: Proceedings of 1st Institute of Electrical and Electronics Engineers (IEEE) International Smart Cities Conference 2015 (ISC2-2015). Institute of Electrical and Electronics Engineers, Guadalajara, Mexico (2015). https://doi.org/10.1109/isc2.2015.7366196. https://rgu-repository.worktribe.com/output/1239044/using-gamification-to-incentivize-sustainable-urban-mobility
14. ÖVUS: 5,3 Millionen Österreicher spielen Videospiele. Tech. Rep. Wien (2019). https://www.ovus.at/news/ueber-fuenf-millionen-oesterreicher-spielen-videospiele/
15. Pfeiffer, A., Bezzina, S., Koenig, N., Kriglstein, S.: Beyond classical gamification: in- and around-game gamification for education. In: 19th European Conference on e-Learning (ECEL), pp. 18–30. Berlin (2020)
16. Pfeiffer, A., Kriglstein, S., Wernbacher, T.: Blockchain technologies and games: a proper match? In: International Conference on the Foundations of Digital Games, vol. 71, pp. 1–4. Association for Computing Machinery, New York, NY (2020). https://doi.org/10.1145/3402942.3402996
17. Schmidt, R.R., Geyer, R., Lucas, P.: The barriers to waste heat recovery and how to overcome them? Discussion Paper. Tech. Rep. Austrian Institute of Technology (2020). https://www.euroheat.org/wp-content/uploads/2020/06/Discussion.pdf
18. Thiel, S.K., Lehner, U.: Exploring the effects of game elements in m-participation. In: Proceedings of the 2015 British HCI Conference, British HCI '15, pp. 65–73. Association for Computing Machinery, New York, NY (2015). https://doi.org/10.1145/2783446.2783587

19. Vorderer, P., Klimmt, C., Ritterfeld, U.: Enjoyment: at the heart of media entertainment. Commun. Theory **14**(4), 388–408 (2004). https://doi.org/10.1111/j.1468-2885.2004.tb00321.x
20. Yee, N.: Motivations for play in online games. CyberPsychol. Behav. **9**(6), 772–775 (2006). https://doi.org/10.1089/cpb.2006.9.772. https://www.liebertpub.com/doi/10.1089/cpb.2006.9.772

"Blockchain May Automate Jobs Done by the Boss and AI Can Predict a Heart Attack"

Discourses of (Imaginary) Applications of New Technologies in Journalism

Tanja Sihvonen, Merja Koskela, and Liisa Kääntä

Abstract This chapter explores how a shared understanding of the potential uses of new, abstract technologies is created in journalism. We analyse Finnish journalistic texts from 2015–2020 in which blockchain and artificial intelligence (AI) are introduced to the readers, conceptualized, and discussed. Drawing on the classic Technology Acceptance Model (TAM), we discuss how journalism frames and reframes the salient attributes of these technologies especially in terms of their perceived usefulness and ease of use. Furthermore, through regarding these speculative meanings as reflections of the *technological* or *algorithmic imaginary* we look into the discourses through which their potential or imaginary applications are explained. This research is important because technologies do not enter the lives of people as concrete devices or programs, but initially as conceptual, imaginary, and affective entities.

Keywords Blockchain · Imaginary · Artificial intelligence · Technology acceptance model · Journalism

1 Introduction

This article explores how a shared understanding of the potential uses of new, abstract technologies is created in journalism. Our focus is on two new technological phenomena, *blockchain* and *artificial intelligence* (AI), and how they are introduced to readers, conceptualized, and discussed. We are interested in the discourses through which these technologies have been defined, and how their potential or imaginary applications have been explained and evaluated (cf. [3, 10]). This is important as technologies do not enter the lives of people as concrete devices or programs, but initially as conceptual, imaginary, and affective entities (e.g. [5]). In this paper, we theorize the issue, present an analysis model, and offer some initial findings through selected examples.

T. Sihvonen (✉) · M. Koskela · L. Kääntä
University of Vaasa, Vaasa, Finland
e-mail: tanja.sihvonen@uwasa.fi

A. Dingli et al. (eds.), *Disruptive Technologies in Media, Arts and Design*, Lecture Notes in Networks and Systems 382, https://doi.org/10.1007/978-3-030-93780-5_3

Technologies, and especially new technological innovations, play a special part in the future orientation of many Western societies. Expectations and promises placed on new technologies are most visible in journalism. While positive framing may generate unrealistic expectations, such as technology hypes, negative framing may constrain or stop the adoption of a technology or delay the acceptance of its applications [30]. Drawing on the classic Technology Acceptance Model (TAM, see [31, 32]), we discuss how journalism frames and reframes (e.g. [25]) the salient attributes of these technologies especially in terms of their perceived usefulness and ease of use. Furthermore, we are interested in speculative meaning-making since we regard meanings as reflections of the *technological* (see [19], p. 67) and *algorithmic imaginary* [4]. These imaginaries reveal something about what kinds of desires for a better society and anxieties for worse are associated with blockchain and AI technologies.

Blockchain and AI are technological megatrends that are thought to have wider social, societal, and cultural impacts. They are expected to change the world around us in unpredictable ways. Blockchain is a technology that solves problems of storing, authenticating, and protecting data through peer-to-peer-operated decentralized ledgers of record. AI as a simulation of human intelligence is based on machine learning, the processing of massive amounts of data, and ever-increasing computing power. It gradually makes it possible for computers to be perceived to interact more like humans, for example to see, read, speak, and even negotiate emotions (e.g. [20, 26]). These technologies hold promise for solving many current problems, and even for easier and better lives for people, but simultaneously, they may come with unforeseen threats.

Hence, the objective of our chapter is to show how the journalistic frames used for representing AI or blockchain in the scenarios studied relate to each other and to the expectancies presented according to the TAM model, and how they indirectly promote the technological and algorithmic imaginaries. Our paper contributes to the earlier discussion on the public perception of new technologies (e.g. [5, 10]), and fills a gap in earlier research by combining two theoretical approaches and by introducing a comparative approach between blockchain and AI technologies. When it comes to complex and multifaceted issues such as these, journalism is an important channel for informing the citizens. It does not only offer an arena for deliberating the threats and possibilities of complicated societal questions, but it also diffuses information about these to people who are not necessarily interested in them. It is important to study how new technologies are framed for different audiences, lay people and specialized publics alike.

2 Theoretical Framework

The concepts of *technological imaginary* (see [19], p. 67) and *algorithmic imaginary* [4] can be used as a framework for understanding the role of imagination in interpreting what new technologies mean for different people. These concepts have been used mainly for the study of new media forms (e.g. [21]) but they also lend themselves to broader applications. According to this framework, cultures tend to attach myths, attitudes, and values to new technologies, in particular if the technologies in question have a potential for making people's lives and society better, or the opposite, for the destruction of social cohesion [13, 19]. As Ferrari states, technological imaginaries are "sets of practice-based beliefs, individual and collective, implicit and explicit, about the role of technology in social life and social change" [11]. She found three types of imaginaries in political social movements discourses: imaginaries of appreciation, negotiation, and challenge. These results relate to our assumption of technological imaginaries to be inherently dichotomic or balancing in nature, such as *good/bad, hopes/concerns, optimistic/pessimistic* [22].

When it comes to blockchain, the technological imaginary today is evident in the narratives of people investing in cryptocurrencies and losing all their assets or getting fabulously rich. It is also present in the dual business discourse focusing either on the disruptive character of technology, e.g., undermining the banking system, or its enormous business potential. The concept of artificial intelligence, on the other hand, is connected to the ideas of people becoming liberated from routine tasks and certain responsibilities in working life, as well as having more free time, but simultaneously it is associated with discourses of increasing unemployment and social injustice ([4], see also [22]).

The concrete way imaginaries and discourses are mobilized in journalistic texts is through the practice of *framing*. Framing is a traditional way of interpreting both the production and consumption of journalistic texts (e.g. [24]). In principle, frames are cognitive structures that help people to interpret what is going on in the world [14]. The concept of 'frame' has been used in many ways in earlier research; frames can be found in texts and in journalists' minds, as well as in the readers' or audiences' interpretive practices, or in culture more generally ([8], p. 53). When it comes to framing a new technology in journalism, relevant frames are media frames that can be located in text based on certain framing devices, such as metaphors or definitions (on metaphors, see [12]). These steer the readers' interpretations. There are also audience frames that readers use to understand and make sense of what they are reading. However, in this article we focus on the ways in which journalists frame new technologies and render them intelligible to their audiences (see e.g. [6]). In practice, this framing is interpreted through what we call *scenarios*, or illustrations of ideas in concrete ways [17].

The technology acceptance model maps factors that influence how a new technology is adopted by its users [32]. These factors are *performance expectancy*, which describes how useful the technology is seen to make work (typically) more efficient; *effort expectancy*, which describes how easy the technology is to use and to understand; *social influence* meaning to which degree important others believe it should be adopted; and finally *the conditions* that facilitate the acceptance of the technology in question. Moreover, several additional factors have an impact on the acceptance rate of new technologies, such as gender, age, previous experience, social context, and whether the use is obligatory or voluntary.

While TAM focuses on how individuals react and behave in new circumstances, interpretations of how journalism constructs new technologies are social and contextual, and based on shared understandings. Therefore, we have chosen to combine TAM as a theoretical framework with other theoretical concepts, the technological and algorithmic imaginary, and framing. In addition, TAM has its limitations (see [1]) in predicting how people behave and in taking into account institutional constraints, but when supported by other conceptual frameworks it can be used as a general background for understanding the discourses of technology in journalism.

3 Data and Methods

In this article, we analyze Finnish journalistic texts from 2015–2020, representing both general and specialized journalism in well-established large newspapers and more experimental online zines. The data was collected from the websites of Helsingin Sanomat (HS), DigiToday, TiVi, Kauppalehti (KL), and Taloussanomat (TS) using the customized search tool and the Finnish search words for *blockchain* ('lohkoketju') and *artificial intelligence* ('tekoäly', 'keinoäly'). The topics of analyzed texts cover various societal themes from financial and technological to social and cultural issues. In particular, business perspectives are prevalent in the data, which is a direct consequence of us including financial journalism in the research material (KL, TS). Originally, the number of search results was very high, which is why we have focused on texts in which blockchain and/or AI are in themselves, or are closely associated with, the main topic. As both blockchain and AI are abstract technologies, the actual topic of these articles is usually something more concrete, such as distinct applications, robotics, etc. As a result of these exclusions, our research data contains 600 articles. In this chapter, we report an initial analysis where we have manually processed the texts until we have reached a sample saturation (approximately at the level of 300 texts).

Our analysis proceeded as follows: First, we employed content analysis (see [2]) to locate text passages where blockchain or AI were described, explained or defined. Following that, we continued the content analysis by extracting scenarios, that is, illustrations of complex and abstract ideas in easily understandable ways (see [15]). The point of using scenarios is that they seem to be the common way of conveying the technological or algorithmic imaginary. Moreover, they form practically sized units of analysis. Scenarios are regularly connected with certain linguistic cues, such as *may be used, can in the future, enables, is planned, is intended for, promise, possibilities,* which make them recognizable and searchable thus allowing larger datasets to be used and even carry potential for corpus-based approaches in the future.

Third, based on the identified scenarios we applied frame analysis [8, 14] to explore which aspects of reality were highlighted to create shared meanings of these new technologies. We understand frames as culturally constructed, and in the data, they are employed to explain the phenomenon at hand in ways that reflect the norms, values, and shared beliefs in society [9, 12]. As the texts in our data demonstrate a reason and an agenda for why journalists have chosen to discuss and represent these technologies in the first place, it is likely that they re-apply pre-existing interpretations to new issues rather than construct new ones (see [29]). We argue that the *expectancies* presented in TAM are such pre-existing lenses. Therefore, in this chapter we discuss how the frames identified relate to the increased efficiency of work (*performance expectancy*), the ease or difficulty of use and understanding (*effort expectancy*), references to "big players" and pioneers (*social influence*), and facilitating factors (*conditions*).

In our analysis we highlight the most common *framing devices*, such as metaphors, exemplification, and concretization [15, 16, 18, 23] as a way of approaching the technological imaginary presented in the scenarios. These devices tend to overlap and co-occur, which makes them difficult to quantify. In the context of the imaginary, *concretization*, i.e. presenting abstract information in a non-abstract manner, is particularly interesting because there lies an initial paradox in the assumption that something is imagined and concrete at the same time.

4 Analysis and Results

In this section, we go through the procedure of analysis and the development of our thinking through selected examples. These examples consist of our own translations of the original Finnish journalistic texts. The examples highlight how our analysis works as a tool for understanding how journalism introduces new technologies to the public and presents them more or less speculatively as useful but still risky ideas. Simultaneously, our discussion shows how the concept of technological imaginary can be used to trace certain discourses in principally fact-based and objective journalistic texts (see [22]). Our qualitative analysis shows how the journalistic frames used for representing blockchain or AI in the scenarios studied relate to each other, to the expectancies presented according to the TAM model, and to the balancing imaginaries of these technologies' significance in society. We demonstrate in detail how the choices of words and other signifiers relate to the scenarios we find particularly interesting.

Our findings reveal that there are two prevailing frames of introducing new technologies to the reader. These are *the frame of opportunities* and *the frame of risks*, which are often combined in the same text possibly guided by the journalistic ideal of impartiality and fairness (e.g. [27]). However, either of the frames may be emphasized while the other one is given a minor role. When it comes to blockchain, the main frame is most often the frame of opportunities: in these texts, a potential problem is presented first and blockchain is offered as a solution. Only when blockchain-based cryptocurrencies are discussed, the frame of risks is highlighted. With AI, the frame of risks seems to be highlighted, even though both frames often co-exist simultaneously.

In the following section, we discuss two examples of scenarios presenting opportunities (scenarios 1 to 2) and two examples of scenarios presenting risks (scenarios 3 to 4) using the analytical concepts presented in the theoretical framework above.

4.1 Scenario 1: Opportunities of Blockchain

Example (1) is recognizable as a scenario of the opportunities of blockchain based on certain choices of words: the modal verb form '*can*' signals the future ability of businesses, and a solution *will allow people to self-authorize* later. An even clearer signal is the use of the phrase *offers … the opportunity* combined with an example. These signals indicate that we have a scenario where technological imaginary is produced: a more or less fictional possibility of a future use of the technology is presented.

(1) Gem [an IT startup] is focused on developing a platform **on which companies can build their own blockchain applications**. [–] Tieto is currently developing a technology that **will allow people to self-authorize the use of their genetic information in medical research**. At this point, the technology is piloted only for consent, and blockchain itself does not store patient data. The technology **offers researchers the opportunity to obtain, for example, the consent of a person carrying a particular gene** to combine genetic data and other health data in an anonymized form. (TIVi 9.11.2017)

The frame of opportunity in example (1) allows for traces of all the expectancies listed in TAM in the scenario. From the point of view of an individual, *performance expectancy* refers to how much using a system is perceived as helpful for attaining a gain in a work task ([33], p. 447–448). Such an *imaginary* gain is offered for companies in the example. A future application is in development and it offers a potential gain. Also researchers are offered a gain thanks to this new technology. Promises of *effort expectancy* in the example indicate that the future application would be easy to use. The phrase "a technology that will allow people to self-authorize the use" hints that people *themselves* will be able to use it.

Additionally, it is suggested that less effort than today will be needed from researchers thanks to the new technology. The scenario thus is that the application provides companies, individuals, and researcher's usefulness and ease of use. In TAM, *social influence* is an external factor which influences the attitudes of users. In the example, naming important developers such as the Finnish company *Tieto* may function positively as social influence as does the high status of medical research. Another external factor is the context of health care, where sensitive personal information is handled. In example (1), the potential risks of combining genetic data with healthcare data of a single person are not highlighted. Rather, they are mitigated by allowing people themselves to give the authorization and also by limiting the application to the authorization phase (cf. [34]).

Thus the scenario of the opportunities of blockchain seems to serve firms, individuals, and researchers equally by promising to make their lives easier and offering them potential gains and ease of performance in the future. Example (1) paints a picture of a positive future where blockchain guarantees the ownership of healthcare data to the individuals rather than the system.

4.2 Scenario 2: Opportunities of AI

Similarly to example (1), in example (2) the scenario of opportunities becomes evident in the word choices that emphasize and frame a specific AI technology as a helpful tool. Phrases signaling this include word choices such as: *technology may help, falls may be detected without carrying a device, scientists see a new way of rescuing people.*

(2) MIT in the United States has experimented with looking through walls using radio waves to see people. Now the technology is spiced with artificial intelligence that automatically detects objects. [–] The method uses artificial intelligence to teach wireless devices to recognize human positions and movements even through a wall. [–] RF-Pose is based on a neural network that analyzes radio signals bouncing off humans. This creates dynamic stick figures that mime what the person being tracked is doing in real time.

Scientists see here a new way to rescue, for example, earthquake victims from collapsed buildings. At the same time, **technology may help** older people live alone. For example, **falls can be detected** without the need for the occupant to carry any device to call for help. But at the same time, the **potential for more accurate and automated surveillance is also evident**. Especially since, according to researchers, technology not only recognizes movements, but **is also able to correctly identify individuals** 83 times out of 100 **based on movements**. (Digitoday 17.6.2018)

Helpfulness and usefulness of the technology is, however, represented with uncertainty and hedging, through the modal verb *may*. As the technology is still in the test phase and not in actual use, developers, actors in the text, and journalists can only speculate about its opportunities. Opportunities are illustrated through concretization, by real life examples but with some speculation (e.g. *see a new way*). This kind of language use and opportunity framing signals technological imaginary that allows us to see new AI technologies as something good and approachable rather than bad or too abstract (see [10]).

The four expectancies listed in TAM are present also in example (2). Traces of performance expectancy may be found throughout the text, and the technology as such is represented as very efficient (*by combining automatic recognition and AI, the technology can identify individuals correctly 83 times out of 100 through the walls*). However, detection through the wall is against common sense, so it seems that those who gain better performance by using this technology are developers and companies, not individuals. Regarding *effort expectancy*, the example shows that especially older people and earthquake victims could benefit from this technology, but not by using it themselves: it helps *older people to live alone* without special equipment, and offers *a new way to rescue ... earthquake victims from collapsed buildings*. The important other named in the text is MIT and its researchers. Mentioning this kind of big players may help readers to value the technology as positive, as MIT is regarded a very trustworthy institution, especially concerning technology development. Besides the scenario of opportunities, there are also some risks (e.g. *surveillance*) that are represented as conditions to the future use of the technology—or something that needs to be considered. This is in line of, for example, Fast and Horvitz's list of both concerns and hopes of AI [10].

Example (2) paints a positive picture of a new technology that combines radio waves and AI by offering several affirmative aspects on its future use. Yet, the opportunities and the usefulness of the technology is offered to companies rather than individuals in the text. Individuals may indirectly benefit from the use of this technology through major global or national players in specific situations.

4.3 Scenario 3: Risks of Blockchain

Even though blockchain is often presented as a solution rather than a problem or risk, there are also risk scenarios when it is applied to cryptocurrencies such as Bitcoin, which suffer from a bad reputation (e.g. [28]). In example (3), the risk scenario is less future-oriented than the opportunity scenarios. On the contrary, it describes what is currently going on in Bitcoin trading. The scenario is signaled by phrases with negative meanings, such as *already flowed far*, *should be identified*, *even in theory ... only less than seven*, *by no means enough*. There is also an if-structure: *will be particularly interesting if*.

(3) **The frantic trading** with Bitcoin **has already flowed far from** the actual blockchain. [–] IN PRINCIPLE, any transfer of bitcoin to another owner **should be identified** in the **blockchain** that maintains the Bitcoin database. According to the blockchain expert interviewed by HS, Bitcoin's blockchain system **is able to clear even in theory** only less than seven trades per second. **That is by no means enough** at the current pace of trade. **In practice, transactions are mainly settled** within trading venues, and ownership of Bitcoins may not be formally transferred to new owners at all until they are already resold. Many trading venues are still unregulated and there is no guarantee for users that pricing, clearing, and monitoring of transactions will take place in the correct order. During the short history of Bitcoin, trading venues have collapsed or simply disappeared and Bitcoins purchased by users have disappeared with them. **The situation will be particularly interesting if prices really collapse**. (HS 8.12.2017)

While the frame of opportunities could easily be interpreted in terms of TAM, this is more complex with the frame of risks, because after all TAM is a technology **acceptance** and not a rejection model. However, the model may be reversed for understanding the frame of risks, i.e. interpreting the factors that are presented as potential risks. Whereas *performance expectancy* refers to how helpful the technology is supposed to be, its opposite would indicate how risky, dangerous or obstructive it is. In example (3), the risk scenario is connected with the lacking capacity of the technology, and the large number and intensity of transactions (*frantic trading*) taking place leading to non-registered assets being sold again, the marketplaces being non-regulated and vanishing along with the Bitcoins of the clients. Thus *performance expectancy* may be exchanged for *risk potential.* With Bitcoin the losing of all assets is presented as more or less an *imaginary* risk.

Comparably, the promises of *effort expectancy* in the frame of risks turn into describing how difficult and challenging the use of the technology is. However, the technological challenge is not described as something facing the individual buying and selling Bitcoins, but an obstacle for the marketplace. In this way, the risk scenario highlights limitations of blockchain technology. The best the technology can do is perform seven transactions per second, which is not enough for the Bitcoin market. The risk scenario develops in a chain of reasons and consequences, where effort expectancy turns into a fear of losses.

In example (3), social influence is represented by unnamed blockchain experts cited by the newspaper itself, or by predictions published in the international business press (Bloomberg). Both of these authoritative voices warn of potential disruption. Unlike the opportunity scenarios, the risk scenario does not offer any mitigating details at the end that would hint at an alternative course of events potentially reducing the risk. In that way, the risk scenario, which is imaginary in the same way as the opportunities are, is represented as non-imaginary and a rather probable consequence of the technology's limitations and the Bitcoin frenzy.

4.4 Scenario 4: Risks of AI

AI mainly appears framed with risk scenarios in our data, possibly because AI remains abstract to most non-expert readers (see [7, 10]). Example (4) is from a business journal, but the topic of the text is entertainment. The scenario of risks is signalled by (1) a description of the spread of the app (*downloaded by more than 100 million people*), (2) a negative characterization (e.g. *almost unlimited and irrevocable rights to use images, its data may be transferred* to locations that have different legislation), and (3) potential negative consequences of its use (*may be used for roadside ads, advertising and surveillance*). We can see that the scenario is framed by combining hedging and future-oriented words with strong adjectives and adverbs such as *unlimited* and *irrevocable*, and a metaphor (*spreading like wildfire*).

(4) Facebook is filled with funny pictures where a person suddenly seems to have aged a lot. [–] The growing old effect is due to an application called FaceApp. According to Forbes, more than 100 million people have downloaded the app to their device. The app is owned by Russian Wireless Labs and it was launched as early as 2017. Now it is spreading like wildfire. Few users of the application have read the user agreement. In it, the user **grants the application almost unlimited and irrevocable rights to use the images**. In addition, FaceApp has access to the person's location. The consumer is also reminded that **the data collected may be transferred outside the EU or the US**, i.e. to Russia, for example, where completely different data protection laws exist. Every user who uploads their images to the app thus gives FaceApp **an unlimited right to do almost anything with these images**. According to Peter Kostadinov, editor of the PhonArena website, the user's **image may end up in a roadside advertisement** in Moscow, for example. However, according to Kostadinov, images are more likely to be used when a company trains artificial intelligence for facial recognition. This **technology can be used, for example, to target advertising or to supervise people**. In addition to photos, also other information about the user is transferred to FaceApp. FaceApp announces that it collects information about, among other things, browser history [–]. (Kauppalehti 18.7.2019)

On the basis of TAM and our extension to it, it is important to see how technology is presented as disruptive or non-essential. In example (4), the technology is described as very efficient, yet entertaining; hence, its usefulness is not speculated in relation to better work or business possibilities, unlike with many other AI technologies. Furthermore, the text frames the app as a potential source of mistakes if its risks are not understood, and if the rights and permissions document is not considered carefully (*effort expectancy*). The text offers knowledge about the *conditions* that contextualize the use of the app which facilitate its rejection; the app is associated with so many risks that individuals should not use it, not even for entertainment purposes.

One major element in this example is that it reveals a group of actors in voicing warnings (*Forbes, specialized journalists*) while others are to be protected (*consumers, individuals*) against the possible misuses of the app's collecting of individual data and images. Therefore, example (4) presents a negative picture of AI. The risks are speculated upon especially from the point of view of individuals, i.e. lay persons who do not benefit from using the app—on the contrary. The usefulness of the app is very limited to an individual, but to the company its potential is presented to be huge. The technological imaginary shows specific AI technologies for entertainment purposes as something you should avoid and reject, or at least consider very carefully before using. In this sense, our data reflects the "fear of loss of control of AI" discussed by Fast and Horvitz ([10], p. 968).

5 Discussion and Conclusions

In this chapter, we have presented an initial analysis model for studying how journalistic texts have framed and reframed two new technologies, blockchain and artificial intelligence (AI). Specifically speaking, we have tested how the concepts of technological and algorithmic imaginary and the Technology Acceptance Model (TAM) allow us to do this kind of analysis. The analysis is qualitative and focused on a limited dataset. Naturally, if these kinds of data were quantified through coding, we would most likely obtain a different view on the framing. Nevertheless, some initial findings can be discussed.

At a broad level, our results indicate that two conventional frames, the frame of opportunity and the frame of risks, are repeatedly used in journalistic texts discussing new technologies. This is a finding that confirms previous studies on similar themes (e.g. [10]). However, our data show that there are differences between how AI and blockchain are represented, and how pros and cons are balanced with each other in journalistic framing. While the negative perspectives and fears of AI are highlighted, blockchain as a newer technology is presented mainly through the frame of opportunities, as long as Bitcoin (or other cryptocurrencies) is not discussed. Therefore our contribution to this discussion lies in the comparative approach and in highlighting the journalistic framing.

From the point of view of making new technologies easy to understand for the general public, our study reveals the use of different popularizing devices (see [18]). The common frames, opportunities and risks, are indicative of the need to make these types of texts interesting to readers, for instance, through presenting them as engaging stories [35] or recognizable metaphors [12]. It is one of the starting points of popularization that journalists would rather paint with a broad brush the future applications of the new ideas than describe them in detail (e.g. [18]). This is illustrated by the scenarios we have analyzed: different means of concretization, examples, and metaphors are used to make the content accessible. This is how the technological and algorithmic imaginary is reproduced in journalism—it is an illustration technique for popularizing science or economic systems. In other words, journalists interpret and illustrate technology-specific knowledge [16] keeping in mind their target audiences.

It is evident that different tools for framing are used for different publics: a lay audience needs more clarification and concretization than business professionals, whose interests may journalistically be served by painting broad business opportunities or alarming warning signals. Interestingly, concretization seems to work at different levels. For example, a lay audience may need concrete descriptions of technologies and their potential uses, but if the technology itself remains abstract or a future imaginary more than anything else, journalists lack ways of concretizing the discussed issue in understandable terms. This reflects the representational and framing nature of media texts. Technological imaginaries are occasionally developed on the basis of sources that either give or gain information about the technologies at hand rather than of functions or uses of specific technologies ('choosing sources', see [3]).

However, it seems that there are broader agendas at play in technology journalism than just educating the public. The scenarios we identified illustrate that the advances of technology and their opportunities are introduced also to serve the specific interests of specialized audiences (cf. [3]). Novel technologies are discussed and explained in order for businesses to be successful in the innovation race, and for decision makers and funders to understand the potential gained through supporting the development of these technologies. Even national interests are often highlighted: a single country is represented as being involved in the innovation race, in competition with other countries.

The other side of the innovation race concerns risks, as both potential and imaginary threats exhibit news potential and engage readers. Presenting risks may have educational goals (such as "do not believe in getting rich quickly by buying cryptocurrencies"), but it also protects prevailing societal institutions and structures, such as banks and technology companies with their specific business interests. The focus on risks in journalism is therefore associated with the maintenance of status quo in society. This finding ties in with Natale's [21] notion of speculations in media representations being cultural visions and imaginaries with political potential. These representations can be explored as efforts to affect how readers see technological changes in society.

In conclusion, we emphasize that technological and algorithmic imaginaries can be seen as balancing devices that simultaneously both guide readers in sustaining their meaning-making regarding emerging technologies, and challenge them to see those technologies more contextually and societally [7, 10, 11]. All technologies have positive and negative aspects. By relying on and framing this balancing activity, journalism perpetuates the instability of technological discourse rather than displays it as something already set. Furthermore, it seems that through this ongoing technological discourse media and journalism guide people to envisioning and *imagining* a future where technologies play a pivotal part, but the questions of control and power are mostly left unanswered.

References

1. Ajibade, P.: Technology acceptance model limitations and criticisms: exploring the practical applications and use in technology-related studies, mixed-method, and qualitative researches. Library Philosophy and Practice (2018)
2. Braun, V., Clarke, V.: Using thematic analysis in psychology. Qual. Res. Psychol. **3**(2), 77–101 (2006). https://doi.org/10.1191/1478088706qp063oa
3. Brennen, J.S., Howard, P.N., Nielsen, R.K.: What to expect when you're expecting robots: futures, expectations, and pseudo-artificial general intelligence in UK news. Journalism (2020). https://doi.org/10.1177/1464884920947535.Publisher:SAGEPublications
4. Bucher, T.: The algorithmic imaginary: exploring the ordinary affects of Facebook algorithms. Inf. Commun. Soc. **20**(1), 30–44 (2017). https://doi.org/10.1080/1369118X.2016.1154086
5. Cave, S., Coughlan, K., Dihal, K.: Scary Robots": examining public responses to AI. In: Proceedings of the 2019 AAAI/ACM Conference on AI, Ethics, and Society, AIES '19, pp. 331–337. Association for Computing Machinery, New York, NY, USA (2019). https://doi.org/10.1145/3306618.3314232

6. Chuan, C.H., Tsai, W.H.S., Cho, S.Y.: Framing artificial intelligence in American newspapers. In: Proceedings of the 2019 AAAI/ACM Conference on AI, Ethics, and Society, AIES '19, pp. 339–344. Association for Computing Machinery, New York, NY, USA (2019). https://doi.org/10.1145/3306618.3314285
7. Elish, M.C., boyd, d.: Situating methods in the magic of big data and AI. Commun. Monogr. **85**(1), 57–80 (2018). https://doi.org/10.1080/03637751.2017.1375130
8. Entman, R.M.: Framing: toward clarification of a fractured paradigm. J. Commun. **43**(4), 51–58 (1993). https://doi.org/10.1111/j.1460-2466.1993.tb01304.x
9. Eyck, T.A.T.: The media and public opinion on genetics and biotechnology: mirrors, windows, or walls? Public Underst. Sci. **14**(3), 305–316 (2005). https://doi.org/10.1177/0963662505052888
10. Fast, E., Horvitz, E.: Long-term trends in the public perception of artificial intelligence. In: Proceedings of the Thirty-First AAAI Conference on Artificial Intelligence, AAAI'17, pp. 963–969. AAAI Press, San Francisco, California, USA (2017)
11. Ferrari, E.: The technological imaginaries of social movements: the discursive dimension of communication technology and the fight for social justice. Publicly Accessible Penn Dissertations, University of Pennsylvania, Philadelphia (2019). https://repository.upenn.edu/edissertations/3513
12. French, M., Hancock, J.: What's the Folk Theory? Reasoning About Cyber-Social Systems. SSRN Scholarly Paper ID 2910571, Social Science Research Network, Rochester, NY (2017). https://doi.org/10.2139/ssrn.2910571
13. Fry, K.: Media literacy education: harnessing the technological imaginary. J. Media Literacy Educ. **3**(1) (2013)
14. Goffman, E.: Frame Analysis: An Essay on the Organization of Experience. Harper & Row, New York, NY, USA (1974)
15. Gülich, E.: conversational techniques used in transferring knowledge between medical experts and non-experts. Discourse Stud. **5**(2), 235–263 (2003). https://doi.org/10.1177/1461445603005002005
16. Kaplan, A., Haenlein, M.: Siri, Siri, in my hand: who's the fairest in the land? On the interpretations, illustrations, and implications of artificial intelligence. Bus. Horiz. **62**(1), 15–25 (2019). https://doi.org/10.1016/j.bushor.2018.08.004
17. Kohring, M., Matthes, J.: The face(t)s of biotech in the nineties: how the German press framed modern biotechnology. Public Underst. Sci. **11**(2), 143–154 (2002). https://doi.org/10.1088/0963-6625/11/2/304
18. Koskela, M.: Ways of representing specialized knowledge in Finnish and Swedish science Journalism. LSP Prof. Commun. **2**(1), 27–40 (2002). https://rauli.cbs.dk/index.php/LSP/article/view/1942
19. Lister, M., Dovey, J., Giddings, S., Kelly, K., Grant, I.: New Media: A Critical Introduction, 2 edn. Taylor & Francis (2009)
20. Marr, B.: 9 technology mega trends that will change the world (2019). https://bernardmarr.com/9-technology-mega-trends-that-will-change-the-world/
21. Natale, S.: Introduction: new media and the imagination of the future. J. Mobile Media **8**(2) (2014)
22. Natale, S., Ballatore, A.: Imagining the thinking machine: technological myths and the rise of artificial intelligence. Convergence **26**(1), 3–18 (2020). https://doi.org/10.1177/1354856517715164
23. Niederhauser, J.: Wissenschaftssprache und populärwissenschaftliche Vermittlung. Gunter Narr Verlag, Tübingen (1999)
24. Pan, Z., Kosicki, G.M.: Framing analysis: an approach to news discourse. Polit. Commun. **10**(1), 55–75 (1993). https://doi.org/10.1080/10584609.1993.9962963
25. Reese, S.: Prologue-framing public life: a bridging model for media research. In: Framing Public Life. Perspectives on Media and Our Understanding of the Social World, pp. 7–31. Routledge, New York, NY, USA (2003)

26. Riedl, M.O.: Human-centered artificial intelligence and machine learning. Human Behav. Emerg. Technol. **1**(1), 33–36 (2019). https://doi.org/10.1002/hbe2.117
27. Sambrook, R.: Delivering trust: impartiality and objectivity in the digital age. Tech. rep., Reuters Institute for the Study of Journalism, University of Oxford (2012). https://reutersinstitute.politics.ox.ac.uk/our-research/delivering-trust-impartiality-and-objectivity-digital-age
28. Swartz, L.: What was bitcoin, what will it be? The techno-economic imaginaries of a new money technology. Cult. Stud. **32**(4), 623–650 (2018). https://doi.org/10.1080/09502386.2017.1416420
29. van Gorp, B.: The constructionist approach to framing: bringing culture back in. J. Commun. **57**(1), 60–78 (2007). https://doi.org/10.1111/j.0021-9916.2007.00329.x
30. van Lente, H., Spitters, C., Peine, A.: Comparing technological hype cycles: towards a theory. Technol. Forecast. Soc. Chang. **80**(8), 1615–1628 (2013). https://doi.org/10.1016/j.techfore.2012.12.004
31. Venkatesh, V., Bala, H.: Technology acceptance model 3 and a research agenda on interventions. Decis. Sci. **39**(2), 273–315 (2008). https://doi.org/10.1111/j.1540-5915.2008.00192.x
32. Venkatesh, V., Davis, F., Morris, M.G.: Dead or alive? The development, trajectory and future of technology adoption research. J. Assoc. Inf. Syst. **8**(4) (2007). https://doi.org/10.17705/1jais.00120
33. Venkatesh, V., Morris, M.G., Davis, G.B., Davis, F.D.: User acceptance of information technology: toward a unified view. MIS Q. **27**(3), 425–478 (2003)
34. Woodall, A., Ringel, S.: Blockchain archival discourse: trust and the imaginaries of digital preservation. New Media Soc. **22**(12), 2200–2217 (2020). https://doi.org/10.1177/1461444819888756
35. Złotowski, J., Proudfoot, D., Yogeeswaran, K., Bartneck, C.: Anthropomorphism: opportunities and challenges in human-robot interaction. Int. J. Soc. Robot. **7**(3), 347–360 (2015). https://doi.org/10.1007/s12369-014-0267-6

Gallery Defender: Integration of Blockchain Technologies into a Serious Game for Assessment: A Guideline for Further Developments

André Thomas and Alexander Pfeiffer

Abstract Blockchain technologies are already being used in many different applications to secure specific data sets. For the same purpose, blockchain is also used in digital games. This technology is not only discussed, but also actively implemented in the educational sector, for example, in such projects as "Blockcerts". Typical applications are issuing certificates and their validation by using a one-way hash; more creative uses of blockchain beyond that are still uncharted territory. This article shows how blockchain (Ardor and its childchain Ignis) has been integrated into a serious game for assessment of students' knowledge. More specific details include encrypted meta-data in the respective transactions, and offer guidelines for future projects in education.

Keywords Blockchain · Blockchain games · Gamification · Serious games · Student assessment

1 Introduction

Blockchain offers significant opportunities, particularly from the socioeconomic perspective, that go beyond the current ways in which we deal with data storage. This includes such areas as art [17], museums [13] and learning. A summary of opportunities of blockchain for education can be found in the report by Grech and Camilleri [7]. Transfer of data sets into the blockchain and potentially rapid verification of their validity opens up new infrastructural and other possibilities that are related to the defining qualities of blockchain-based technical solutions [7, 11].

A. Thomas (✉)
Texas A&M University and Triseum LLC, College Station, TX, USA
e-mail: manink@arch.tamu.edu

A. Pfeiffer
Massachusetts Institute of Technology, Cambridge, MA, USA
e-mail: pfeiffer@alexpfeiffer.at

A. Dingli et al. (eds.), *Disruptive Technologies in Media, Arts and Design*, Lecture Notes in Networks and Systems 382, https://doi.org/10.1007/978-3-030-93780-5_4

- **Transparency** enables provenance of digital assets, i.e. artworks, so their buyers and sellers can conduct transactions and be sure that each party has the right to enter into that transaction;
- **Controllability** (self-sovereignty) allows individuals to identify themselves online while at the same time maintaining control over their personal data;
- **Trust**, in its algorithmically ensured form, provides enough confidence in payments and verified certificates;
- **Disintermediation** removes central controlling authority from the interactions between different blockchain users: the transactions are carried out in a peer to peer fashion, and the required records are made available to every participant;
- **Immutability** of a decentralized ledger of records enables permanent storage, without the possibility of modification unless designed otherwise;
- **Collaboration** can be carried out in a more democratic manner, as the parties can transact directly with each other without the need for mediating third parties.

Based on the information available in 2017, the report concludes that the use of blockchain applications in education is still in its infancy, although it is gaining fast momentum. A number of case studies have been conducted nevertheless, including the implementations at the Open University UK, the University of Nicosia, MIT and various educational institutions in Malta: each of these implementations were in a pilot phase at the time of the report [7]. The results of the pilot phase suggest that blockchain technologies might have a disruptive effect on current providers in the educational software market in the near future. As of recently, this disruption is becoming the reality, as many educational institutions and services has moved online due to the pandemic situation in 2020/21.

Based on [7], the following use cases for blockchain solutions can be named in relation to education.

- Creation of digital certificates and credentials, as well as designing digital proofs of authenticity for material certificates and printed credentials;
- Immutable storage of proofs of performance at examinations, including the related metadata;
- Recognition of examination results between and within educational institutions without the need for the central authority;
- Creating a personal "lifelong learning" record of achievements (a virtual CV) on blockchain;
- Verification of authenticity of the certificates issued by third parties;
- Management of intellectual property, e.g. results of research in the context of project implementation;
- Processing of payments, etc.

The demonstrator presented in this paper builds on Grech and Camilleri's vision of the rapidly approaching future regarding blockchain in education [7]. The aim of the team of authors was to develop a serious game [16], complete with the option for examination that transmits the results of the tests to all relevant persons (teachers as well as students). This solution is designed to include all relevant metadata, such as the curriculum stored in the database, the issuing institution, selected personal details of the examiner and examinee, etc. into transactions on blockchain.

The research that accompanies the demonstrator addresses the following research question:

> Which basic requirements are needed to use blockchain technologies not only for the verification of original documents by using a one-way hash, but to use blockchain directly with assessment instruments in real-time environments, taking technological, social and commercial factors into account?

2 Related Work

The last several years have brought considerable development in the area of blockchain research in the field of digital games, specifically in the field of education. However, there is hardly any or even no publication on serious games in connection with blockchain technologies. Below we look at the most notable working papers and articles that deal with the subject of distributed ledgers and games in the ways that can be applied to gamification of education.

A number of studies focuses on the design challenges of blockchain-based games. Kraft [10] explores the problem of generating fair randomness in a deterministic, multi-agent context (for instance, a decentralized game built on a blockchain). He states that the existing state-of-the art approaches are either susceptible to manipulation if the stakes are high enough or if they are not generally applicable. Therefore he introduces a new approach based on game theory. Min et al. [15] discuss the blockchain integration for games and then categorize existing blockchain games from the aspects of their genres and technical platforms. In practice, Cai et al. [4] develop two different prototypes to research the aspect of multi-chain as well as multi-game approaches with the aim to create an interoperable blockchain gaming framework. Lee et al. [12] examine the effects of speculative and enjoyable aspects of users' playing behavior in blockchain-based collectible games. As collectibles require dedicated marketplaces to circulate, Chen [5] observe blockchain technology as an innovative element of the virtual items trading platform. He concludes that the biggest impact of blockchain technology in the short term is the redistribution and integration of the game and trading platform market.

Gamification of education has a relatively small number of researched blockchain cases. In one of the earliest relevant studies, Jirgensons and Kapenieks [8] compare Blockcerts with Ethereum Smart Contracts developed by Open University, UK, while Baldi et al. [3] describe how to impersonate a legitimate issuer of Blockcerts certificates with the aim to produce certificates that cannot be distinguished from originals by the Blockcerts validation procedure. Setting the groundwork for the project presented in this paper, Pfeiffer and König [16] discuss the use of blockchain technologies in educational games for assessment from a humanities perspective. Komiya et al. [9] examine consensus algorithms as a reward system. They introduce their approach "Proof-of-Achievement (PoA)" which is an algorithm optimized for blockchain games, focusing on the number of tasks achieved in the game. The aim of using the PoA algorithm is to increase the motivation for playing blockchain-based games. Similar to Komiya, Yuen et al. [20] also perceive consensus algorithms like Proof of Work "PoW" as a bottleneck for games. Their proposal is called "Proof-of-Play" with a focus on Peer-to-peer (P2P) Games. Agustin et al. [1] describe application of blockchain technology in e-certificates in the open journal system. The study reports that issuance of e-certificates in an open journal system is a way to manage and verify, prevent duplication or even falsification of e-certificates and the reputation of the open journal system is already given. This project is based on Blockcerts by Learning Machine (originally developed at MIT). Finally, Aini et al. [2] explore different approaches to gamification that embed blockchain technologies in the educational sector.

3 The Context of the Demonstrator

To set up the demonstrator in the educational framework, we developed the educational game Gallery Defender on blockchain. This serious game replays the requirements of art history knowledge in the introductory college level Art History Curriculum Framework:

The learning goal equals the game goal [6] and is defined in the following way.

> All students, which equals the players of the game, will understand, analyze, and describe art styles in their historical, social, and cultural contexts.

Gallery Defender is inspired by the game ARTé: Lumiere [19] already available to students online. In our case, the new serious game introduces art concepts to the player/learner at the first stage and assesses the acquired knowledge of them at the following stage. The results are then transferred to the educational institution.

The game is a turn-based card game that consists of multiple levels. It develops a simple narrative, according to which the player/learner is invited to take the role of a gallery owner. Using their profound knowledge of art history, the player must defend the artworks of the gallery from a masterful thief (Fig. 1).

Each level provides the player/learner with a deck of cards that carry images of historical paintings on them. A small subset of the deck is dealt into the hands of players at the start of each turn. Each card corresponds to the action, which is denoted by a small icon; these actions include the attack icon, represented by a "race car" icon, and defense actions, represented by the "headphones", "helmet", and "building" icons. (These actions were chosen from existing icon assets that the authors already owned, and the icons themselves are not related to learning outcomes.)

The player/learner plays against a simple AI who declares an action at the start of each turn. There is exactly one painting in their hand that corresponds to the AI's declared action at any time, which must be correctly matched to the art movement that it belongs to by dragging the card into one of several labeled buckets near the bottom of the screen.

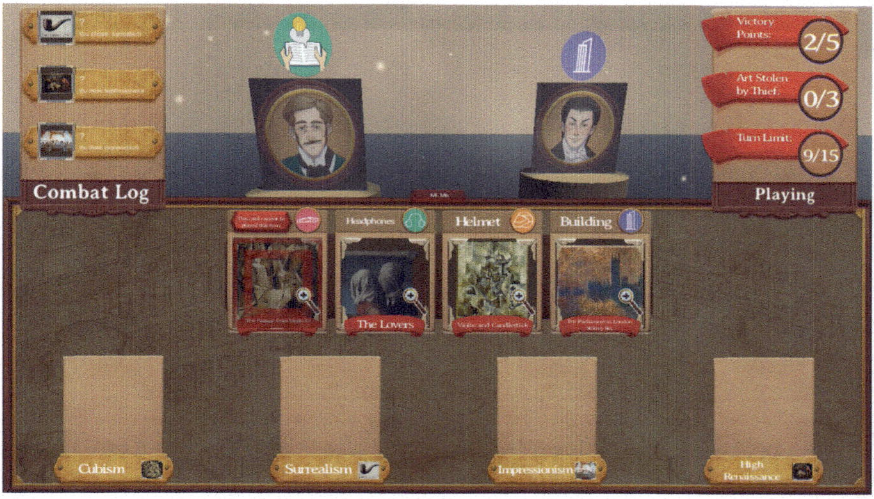

Fig. 1 Screenshot of gameplay in the first version of the game

After successfully matching some number of artworks (which vary by level), the player/learner can move on to the next level. Incorrectly matching too many artworks to art movements will result in failure of the level. Not all levels contain the same number or types of art movements. Additionally, certain levels may allow the player/learner to click on various buttons to gain more information about artworks, art movements, and their previous moves. The Simulation and Assessment Levels are based on the highest difficulty settings possible.

Following the classification of König [16], Gallery Defender is built on the Game Based Learning and Assessment framework. This means that the content and methods are identical in both learning and assessment mode. However, these areas are strictly separated from each other, and the user learns when they step from one sphere into the other.

Gallery Defender utilizes blockchain technologies in three different ways.

The player/learner receives a digital token at the end of the assessment, which contains the grade, points, and time that the assessment was finished as a message. This token and the attached message are stored forever and unchangeably on the blockchain in the player/learner's Ardor Blockchain Wallet. The certificate is encrypted and can only be decrypted by the sender and the original recipient. However, using a shared key enables the player/learner to share the results with third parties, such as their future boss, their family or another school/university (Fig. 2).

The tokens used in the token system of the demonstrator are defined as utility tokens, minted on the Ignis child chain of the Ardor Network. It is of utmost importance that the tokens cannot be traded or sent outside of the learning/assessment environment to prevent speculative behavior [12].

In addition, the player/learner with particularly good results receives an additional token. This token can be exchanged for digital rewards, e.g. a game poster, which

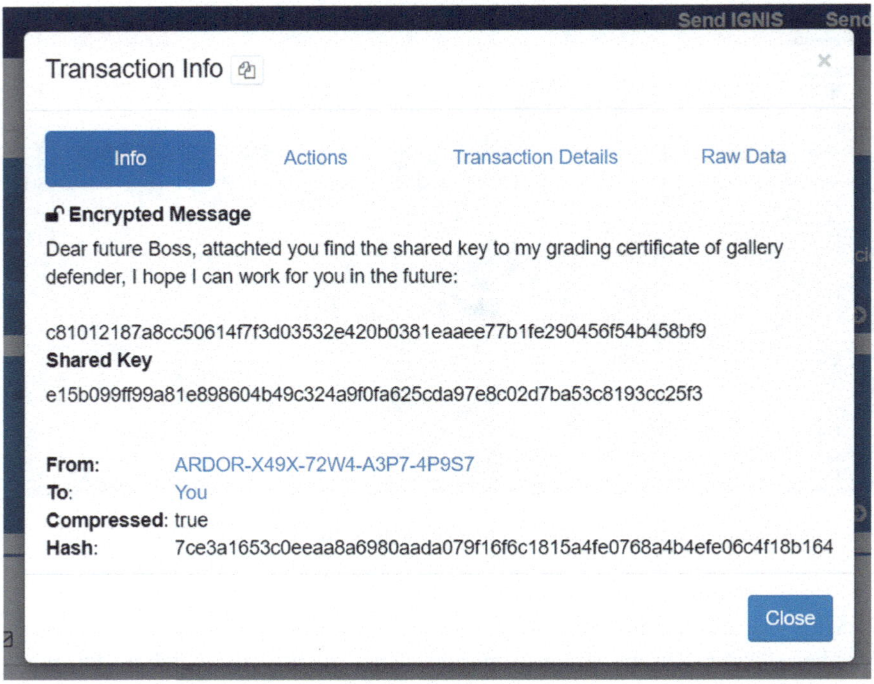

Fig. 2 Screenshot of a message on Ardor blockchain

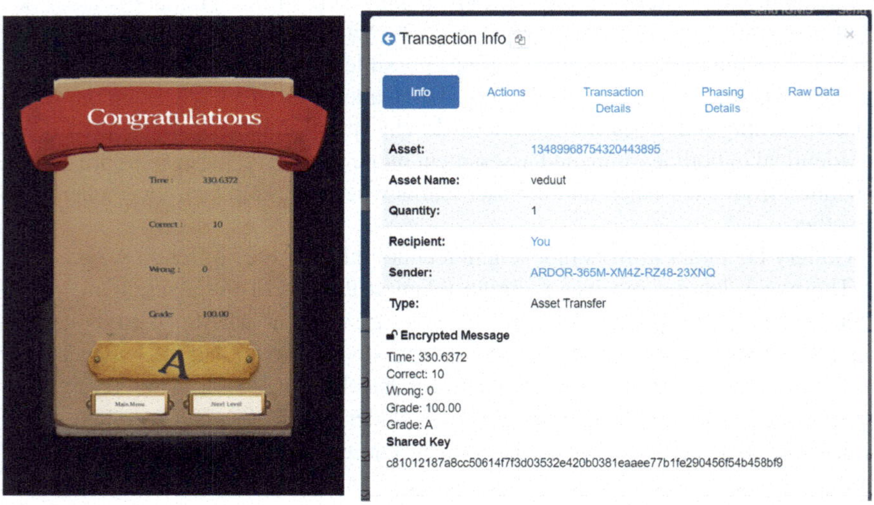

Fig. 3 The grade in the game (on the left) is translated into the exact amount in the blockchain tokens (on the right)

corresponds to the gamification principle realised on blockchain. This functionality can be used to build an entire ecosystem for rewards on blockchain within the context of education, as well as beyond this context.

Finally, the third type of a token is the verification token. Teachers can request this token with further information about the respective test result. After the assessment has been carried out, within a definable time the message is deleted, and only the proof that the token has been sent remains immutable on blockchain (Fig. 3).

4 Methodology

Gallery Defender was created in three iteration steps. The first two iterations took place on the testnet of the system, the third and last iteration was then implemented on the mainnet of the Ardor blockchain. These particular stages were evaluated by using online questionnaires. These questionnaires were filled out at live events such as the Connected Learning Summit, Irvine or the Future and Reality of Gaming Conference, Vienna. Additional participants were recruited, and they received a hyperlink via e-mail to test the demonstrator. The link could be requested using the contact address on the research website of the project. The total number of participants on this survey during the iterative design process was n = 105. The professional background of the survey participants ranges from education and technical professions to people with a background in sociology. This is the case due to the two specific events at which the tests were conducted. A more detailed report on testing has been published [18].

In the first place, testing helped find bugs in the solution. Secondly, new features were suggested by respondents, for example, on how to improve the gaming experience. The focus, however, was on the implementation of blockchain technologies to validate test results from both the socio-economic and technical perspective. For this purpose, open fields could be filled out and therefore both qualitative and quantitative feedback was collected. This was evaluated using the methodology of qualitative content analysis according to Mayring [14] and is discussed in the following section.

5 Results

The results of the content analysis as part of the iterative design process have led to the following guidelines and recommendations.

Either a public blockchain or a combination of private/consortium blockchains should be used in a setting that makes sense and does not limit the reasons why blockchain itself is being used. For instance, in one proposed architecture, personal data is stored using a private blockchain to ensure GDPR compliance. Other highly sensitive data remains in the private area of the respective institution, but the private blockchain sends the hash-value of each block-height to a public blockchain to reduce the possible attack vector for manipulation.

More attention should be paid on the performance of the used networks, regarding number of possible transactions per specified timeframe. In our project, "non freely tradeable utility tokens" are being used (as opposed to fungible and non-fungible freely tradeable tokens). More specifically, the tokens cannot be re-transferred to other users freely, but also, the issuer cannot withdraw them from the user either.

Transaction fees to the network are paid by the respective issuer(s) or operator(s) of the token(s). Therefore the users themselves do not need to hold any "tradeable network maintenance utility tokens", or "cryptocurrencies", nor do they need to know about blockchain. For users the experience should equate to simply using an app. More advanced users can always track their transactions using a public available blockchain tracker or their very own node.

It is possible to encrypt the message associated with the token, so only the transaction itself is publicly visible, while the private information remains private and only accessible to those parties who should have access to it (for the relevant period).

To achieve this shared-keys (or even sub-shared keys, making information available for a specific timeframe or for dedicated users) can be generated, and thus a sharing system for third parties who get access to the information attached to the token can be developed (but not the token itself as information bearer).

The message containing the core content, in certain cases where it makes sense, can be deleted from the blockchain after some time (with the knowledge of the partners involved). Only the proof of the transaction itself remains to reduce the blockchain bloat.

Users can, if they want, always operate a full node and this with existing devices such as their own modern smartphone, their own PC or with inexpensive hardware such as a Raspberry Pi.

The Full Node can be operated without holding network maintenance tokens or cryptocurrencies with the option to run a copy of the blockchain and look into it.

Holding "tradeable network maintenance utility tokens", or "cryptocurrencies" is only necessary if the users want to be part of the active network and conduct mining/forging.

The availability of digital qualified signatures, or other secure forms of digital and forgery-proof identity verification, like an advanced Self-Sovereign Identity System.

A blockchain wallet needs to be created, similar to well-known solutions like the "Exodus Wallet", which can hold different tokens, issued on various systems. However, instead of holding "cryptocurrencies" this time it is a utility token as information carrier.

Further education on the topic of blockchain is needed, if private individuals and companies operate directly with private keys, generated when they open a respective blockchain wallet.

Assuming that the time is not yet there in regards to trusting users to handle a unique key that cannot be restored if lost, it can be defined that the private key itself is only known to application utilizing blockchain to secure certain data, while setting up a classic user and password system which starts the blockchain operations.

Finally, we strongly encourage the designers and adopters of such systems to 'think green': opt-in for solutions that are safe but environmentally friendly. This is especially important for use cases like storing educational data from a game, which do not store monetary records. Therefore, we call for evaluation of a consensus algorithm that fits best for the purpose intended.

6 Conclusion

From these key findings, the following questions can be identified, which researchers, developers and users of systems and applications using blockchain technologies should ask themselves.

- Is a private or public blockchain used?
- Which Blockchain or combination of blockchains systems is used?
- Can the information-carrying tokens be sent from one account to another without this being originally intended?
- Is the private information of the token sufficiently encrypted? And if so, who has the keys to access the data?
- Can (temporary) shared-keys be created that give access to (specific) information for third parties?
- Is it possible for companies, universities and government institutions to operate their own nodes in the network and thus have data sovereignty? If so, how much effort and costs are involved?
- Is it even possible for users to run their own node? And if so, with what effort and costs?
- Do the partners involved have to purchase cryptocurrencies? If so, for what purposes and from which sources? And subsequently, do the users have to purchase cryptocurrencies?
- Who pays the transaction fees to the network? Can the solutions be used that do not charge transaction fees? If this is the case—which network was used and what are the effects of using a transaction fee-free solution?
- Who creates the required blockchain wallets—partners of the respective project or users themselves? Can this be done for them in an automated manner "through the app"? If the latter, who and in what form has control over the private keys?

Although blockchain technologies are constantly evolving from a technological perspective and a lot has been achieved since Grech and Camilleri's report [7], blockchain technologies have not yet arrived at the mainstream areas of application. We could observe this during the content analysis stage of our research: those survey participants who linked blockchain to speculative hype expressed critical opinions on the topic. The second aspect that could be read out of the survey was that there are many blockchain developers and scientists who are fixated on a specific system and do not want to look at other systems or multi-chain solutions.

Therefore, there is still a lot of educational work to be done. One can hope that in the next few years a lot of applied research will be conducted, which will also reach the end users and thus communicate the topic hands-on. A very important finding of the project was also that the demonstrator and the testing together with the proponents, which had little contact with the matter, the active testing helped them to better understand blockchain, as they stated in the follow-up discussions.

7 Future Research

Three aspects will be addressed in the follow-up to the research project presented above.

- Ethical integration of the blockchain solution with digital identity systems.
- Solutions for management of private or shared keys, so that the system is secure but also understandable for its users.
- Further development of interfaces between different parts of the system, so that a wallet can read information from different tokens, and the learning or assessment software can transfer the results to different agents as it is appropriate for the user or the institution.

Acknowledgements The authors express gratitude to Michael Black, Lloyd Donelan and Nick Muniz for their work on the game, to Alexiei Dingli, Vince Vella and Thomas Wernbacher for their invaluable support and guidance, and to Alesha Serada for their editing work and support.

References

1. Agustin, F., Aini, Q., Khoirunisa, A., Nabila, E.A.: Utilization of blockchain technology for management E-certificate open journal system. APTISI Trans. Manag. (ATM) **4**(2), 134–139 (2020). https://doi.org/10.33050/atm.v4i2.1293
2. Aini, Q., Rahardja, U., Khoirunisa, A.: Blockchain technology into gamification on education. Indones. J. Comput. Cybern. Syst. (IJCCS) **14**(2), 147–158 (2020). https://doi.org/10.22146/ijccs.53221
3. Baldi, M., Chiaraluce, F., Kodra, M., Spalazzi, L.: Security analysis of a blockchain-based protocol for the certification of academic credentials. arXiv:1910.04622 (2019). http://arxiv.org/abs/1910.04622
4. Cai, W., Wu, X.: Demo abstract: an interoperable avatar framework across multiple games and blockchains. In: IEEE INFOCOM 2019—IEEE Conference on Computer Communications Workshops (INFOCOM WKSHPS), pp. 967–968 (2019). https://doi.org/10.1109/INFCOMW.2019.8845288
5. Chen, J.T.: Blockchain and the feature of game development. In: Frontier Computing: Theory, Technologies and Applications (FC 2019). Lecture Notes in Electrical Engineering, pp. 1797–1802. Springer, Singapore (2020). https://doi.org/10.1007/978-981-15-3250-4. https://www.springer.com/gp/book/9789811532498
6. Gee, J.P.: What Video Games Have to Teach Us About Learning and Literacy, 2nd edn. Macmillan (2014). Google-Books-ID: v_XIBAAAQBAJ
7. Grech, A., Camilleri, A.F., Inamorato dos Santos, A.: Blockchain in education. Tech. Rep. Publications Office of the European Union (2017)

8. Jirgensons, M., Kapenieks, J.: Blockchain and the future of digital learning credential assessment and management. J. Teach. Educ. Sustain. **20**(1), 145–156 (2018). https://doi.org/10.2478/jtes-2018-0009
9. Komiya, K., Nakajima, T.: Increasing motivation for playing blockchain games using proof-of-achievement algorithm. In: Fang, X. (ed.) HCI in Games. Lecture Notes in Computer Science, pp. 125–140. Springer International Publishing, Cham (2019). https://doi.org/10.1007/978-3-030-22602-2_11
10. Kraft, D.: Game-theoretic randomness for blockchain games. arXiv: 1901.06285 (2019). http://arxiv.org/abs/1901.06285
11. Lapointe, C., Fishbane, L.: The blockchain ethical design framework. Innov. Technol. Gov. Glob. **12**(3–4), 50–71 (2019). https://doi.org/10.1162/inov_a_00275
12. Lee, J., Yoo, B., Jang, M.: Is a blockchain-based game a game for fun, or is it a tool for speculation? An empirical analysis of player behavior in crypokitties. In: The Ecosystem of e-Business: Technologies, Stakeholders, and Connections. Lecture Notes in Business Information Processing, vol. 357, pp. 141–148. Springer (2019)
13. Liddell, F.: Building shared guardianship through blockchain technology and digital museum objects. Mus. Soc. **19**(2), 220–236 (2021). https://doi.org/10.29311/mas.v19i2.3495
14. Mayring, P.: Qualitative Inhaltsanalyse. In: Handbuch Qualitative Forschung in der Psychologie: Band 2: Designs und Verfahren, pp. 495–511. Springer Fachmedien, Wiesbaden (2020). https://doi.org/10.1007/978-3-658-26887-9_52
15. Min, T., Wang, H., Guo, Y., Cai, W.: Blockchain games: a survey. In: 2019 IEEE Conference on Games (CoG), pp. 1–8 (2019). https://doi.org/10.1109/CIG.2019.8848111
16. Pfeiffer, A., Koenig, N.: Blockchain technologies and their impact on game-based education and learning assessment. In: Savegame. Perspektiven der Game Studies, pp. 55–67. Springer VS, Wiesbaden (2019). https://doi.org/10.1007/978-3-658-27395-8_5
17. Scherling, L.S.: Blockchain technologies in community-based arts: implications for fair use and changing practices in art education. Tech. Rep. Columbia University (2017)
18. Thomas, A., Koenig, N., Higgins, T., Black, M., Pfeiffer, A., Donelan, L., Lenzen, B., Muniz, N., Patel, K., Taylan, A., Wernbacher, T.: From learning to assessment, how to utilize Blockchain technologies in gaming environments to secure learning outcomes and test results. MCAST J. Appl. Res. Pract. **3**(2), 172 (2019)
19. Thomas, A., Ramadan, H., Campana, L., Leiderman, D., Sutherland, S., Zawadzki, M.: ARTé: Lumiere (2018). Available electronically at https://oaktrust.library.tamu.edu/handle/1969.1/188003
20. Yuen, H.Y., Wu, F., Cai, W., Chan, H.C., Yan, Q., Leung, V.C.: Proof-of-play: a novel consensus model for blockchain-based peer-to-peer gaming system. In: Proceedings of the 2019 ACM International Symposium on Blockchain and Secure Critical Infrastructure, BSCI '19, pp. 19–28. Association for Computing Machinery, New York, NY (2019). https://doi.org/10.1145/3327960.3332386

Cryptogames as Drivers for Blockchain Application Development

J. Tuomas Harviainen, Alesha Serada, and Tanja Sihvonen

Abstract In this chapter, we outline reasons why cryptogaming, like cryptocurrencies in general, is an area of special interest for blockchain research. We suggest that cryptogaming is a field that drives forward the development of blockchain applications. Games require continuous interest, added value through interaction, and sufficient content in order to stay viable as a business model. They therefore challenge the ways in which we commonly view blockchains as first and foremost trust technologies and sidesteps from traditional banking, contracts and insurance, and call us to envision a much wider application of the technology than is currently being discussed in both popular media and research.

Keywords Cryptogames · Blockchain · Non-fungible tokens · Value propositions

1 Introduction: Serious and Not-So-Serious Applications of Blockchain

Blockchain is widely understood as a universal technological enabler of trust due to the transparency and immutability of data it conveys. There is no need to appeal to a third party as a trustee to verify a transaction between two agents on blockchain, nor supposedly any question about ownership (see e.g., [1]). For this reason, public discourse has labelled blockchain technology as "the trust machine" [2]. Some of the nuance has been lost in this discussion, though: third-party verification is not required in this particular configuration, but the human factor of trust remains. Trust between human actors is crucial in situations such as 'hard forks', operating cryptocurrency exchanges, initial coin offerings, and acknowledgement from traditional financial structures [3, 4].

J. Tuomas Harviainen (✉)
Tampere University, Tampere, Finland
e-mail: tuomas.harviainen@tuni.fi

A. Serada · T. Sihvonen
University of Vaasa, Vaasa, Finland

A. Dingli et al. (eds.), *Disruptive Technologies in Media, Arts and Design*, Lecture Notes in Networks and Systems 382, https://doi.org/10.1007/978-3-030-93780-5_5

Cryptocurrencies have gone a long way from the crypto-anarchic intentions of the original Bitcoin paper [5] to professional trading instruments such as the JPM Coin developed by Morgan [6]. Today, ongoing blockchain development projects aim to enable traceability and proof of ownership to complicated spheres of operations such as land registry [7], identity management [8, 9], and information management in supply chains [10]. There is yet another growing phenomenon within blockchain applications: the so-called cryptogames. For the most part, they are ignored—or at least overlooked—in the otherwise very thorough blockchain research (see [11] for an example). This appears to be because they come from a non-serious direction and are not seen as relevant for societally or financially oriented research on topics such as security tokens, altcoins or smart contracts. However, these "leisures of blockchain" [12] may be significant from a sociological and psychological point of view, and furthermore, important in introducing blockchain applications to the wider public.

In this brief chapter, we outline reasons why cryptogaming, like cryptocurrencies in general, is an area of special interest for blockchain research. Our starting point is the notion that cryptogaming is a field that drives forward the development of blockchain applications, and that games as an area of technological innovation and popular culture is ideal for "public testing" of new technologies. Games require continuous interest, added value through interaction, and sufficient content in order to stay viable as a business model. They therefore challenge the ways in which we commonly view blockchains as first and foremost trust technologies and sidesteps from traditional banking, contracts, and insurance, and call us to envision a much wider application of the technology than is currently being discussed in both popular media and research.

2 Cryptogames

Cryptogames are digital games the designers and publishers of which claim to provide additional value to their users or developers from one or more aspects of the blockchain technology. These aspects, as observed at trade events and on game developers' and games' home pages can be, for instance, permanent ownership of non-fungible tokens (in the form of items, creatures, etc.), transportability of non-fungible tokens to other games, mining of cryptocurrency through play, or the ability to use one's cryptocurrency acquisitions for play instead of spending them on illegal cryptomarkets or taxable currency exchanges.

So far, cryptogames have manifested mainly in the form of casino-type games, collectible-based games, and attempts at recreating the traditional genres of gaming (massively multiplayer online games, first-person shooters) through adding blockchain-based value to them [13]. Similarly to cryptocurrencies, crypto-collectibles "reveal" their value in trading and may eventually become another form of investment. However, the focus on their potential as investments may overshadow their entertainment value (see [14]).

Cryptogames are one of the key innovation areas in the development of blockchain technologies. Their innovativeness is based on the requirement of continuous appeal to users and players, as well as their emerging, more or less functional business models [15]. Most developers of blockchain applications do not have to put efforts on user appeal, because they target very specific audiences or a specialized userbase to begin with. In contrast, a game needs to be playable and interesting for as many people as possible, and especially multiplayer games online depend on the number of players willing to dedicate their time and effort to gameplay. The popularity of games, in general, is constantly jeopardized by play being a voluntary activity, competition from other games and other ways to enjoy leisure, and fluctuating userbases which make the focus on online games a risky business.

Many cryptogames rely on the existence of popular cryptocurrencies or have their own currency as either proofs of work or as added value, supposed or real. One such example is *Axie Infinity* [16], a relatively popular game about training small creatures to engage in combat with each other. These breeding and battling mechanics make it possible to earn a variety of blockchain-based tokens in the game; however, the game runs on Ethereum, and a new player needs to buy a starting set of digital creatures with Ether. Therefore, all players must pay the "entry fee" when joining in, by acquiring the necessary currency, but on the other hand the obligation to use such currency may drive players away even before they try out the game itself [15]. In essence, as the pre-conversion of players into paying customers exists already at the entry level of such cryptogames, the business model of these games needs to be focused on player retention.

Taurion [17] is an example of a game that presents its players with a pervasive online world of planetary exploration and conflict. The company, however, built its public presentations upon the promise that the value of the game would be based on permanent ownership of items via non-fungible tokens, as well as on an integrated alt-coin that the players could earn and use. The question regarding this type of a business model then becomes whether enough people are willing to play the game, as with any MMO (see [18]). Xaya also developed the official *Soccer Manager Elite* for the British game provider. In that game, blockchain is used to manage the ownership of shares in soccer teams and dividends received [19].

For player retention to work as intended for the company, there needs to be a sufficient amount of interesting play for the players of a game (play value). Neither novelty value nor sunken costs are able to keep players interested in the longer run [20]. The challenge for the nascent industry is to keep the game interesting and profitable, while avoiding direct pay-to-win mechanics (i.e., the ability to get an extreme advantage in competitive play by paying real money to the company; [21]). Axie Infinity, too, has raised the question on whether it has been too focused on the pay-to-win (or "invest to win" [22]) mechanic, or whether it is in fact a nicely playable way to get introduced to non-fungible tokens.

The question of sufficient interest and demand remains valid. Outside of certain countries with strict gambling laws, for example, cryptocasinos are unlikely to hold mass appeal. It is easier to speculate with the value of something like Bitcoin than it is to both speculate with its value and simultaneously use it on a virtual roulette. Here we have two significant cases to present: *Flow* and *Terra Virtua*. Flow is a blockchain platform developed by Dapper Labs [23], the company that also owns and updates the so-far most successful cryptogame *CryptoKitties*. In its current state, the description of Flow suggests that it will not be compatible with the Ethereum platform, but it will instead be seamlessly integrated with *Libra*, the ambitious blockchain project of Facebook [24]. As the Google login option will be enabled in Flow, the current speculation is that it will likely exist in partnership with large centralized ad-serving platforms than offer an alternative to them in terms of privacy.

Terra Virtua is another example of a blockchain platform that may have an effect on the game industry in the future. Its goal is to create a mass-market ecosystem for digital collectibles, including exclusive offers from major brands. Terra Virtua utilizes the Unreal Engine from Epic Games to develop its virtual world [25]. Terra Virtua's project, like certain attempts at creating blockchain-based yet sufficiently dynamic first-person shooters, point towards the idea that it may not just be technological diversity and innovation, in the form of playability, but also innovative API's, where we will see the true contributions of cryptogames to blockchain development. Other producers, such as Blocklete Games [26], are in turn forming partnerships with media giants to secure themselves the possibility of including cryptopayments backed by more established businesses. This parallels with earlier cooperation between social media platforms or major media IPs and game producers, such as in the case of Zynga (see [27]). The results of these moves remain to be seen—but are certainly worthy of further research.

3 Discussion

Above we have outlined a number of examples of recent developments, promises and challenges, in the cryptogaming sector. Our argument is that the role of games as part of people's leisurely choices and the voluntary nature of gameplay, as opposed to the utilitarian ethos of adopting and using financial and administrative applications, make cryptogames quite different an area of investigation than other blockchain-based technologies. There is no doubt that due to their nature as mass-appealing and low threshold-emphasizing applications cryptogames drive and accelerate the evolution and diffusion of blockchain technologies. Currencies are often binding, if exchangeable, for their owners. The same thing is true for many other non-fungible tokens, from securities to digital artworks. Games add a wild element into this mix, especially when connected to clever new API solutions.

Games also advance or even force technological innovation in this sector. People often have very strong platform preferences on where, when and with what they want to play. In order to collect the necessary revenue to survive in a financially viable way, cryptogame developers have to adapt to this and push the technical envelope of the entire sector further. If a large potential cryptogame market was, for example, found to exist on a console environment, we would witness significant adaptive leaps in the same way we are now seeing with decentralized apps. Likewise, we believe that the cryptocurrency micropayment market, still in its infancy, will find its strongest developments in this sector.

Cryptogames are still coming into existence and finding their ways of being financially stable. Many cryptogames on the market today are merely copycat versions of previously existing games or variations of early online pet games such as *Neopets* [28], to which blockchain technology is supposed to somehow magically add significant value. However crude in their execution, cryptogames are, in our view, at the core of this topic, if not the discussion. The way in which they rely upon either cryptocurrencies for their value propositions, or on non-fungible tokens for their uniqueness, means that they are increasingly often cutting-edge applications of blockchain technology—just not very popular ones. The question therefore becomes, in their seeking of sufficient player bases, public interest, funding interest, and user retention, where will they take this technology that can be regarded boring by people not interested in market price investment and speculation?

4 Directions for Future Research

In the final section of our chapter, we outline the value of cryptogames to blockchain development and research, as well as point to potential future directions. With the field still being so diverse with regards to game types and application ideas, it is recommendable to seek out interesting examples of successful and not so successful cryptogames for deeper case studies and development inspiration. At the same time, it is essential to keep an eye out on what else is taking place in the cryptogame development as well as game and blockchain industries on the whole. Understanding cryptogames may not only tell us what blockchain is able to do next, but also what it is not able to accomplish. The voluntary nature of gaming, underscored in this brief chapter, creates an interesting stress test for the adoption of new applications, similarly to the manner in which masses of accumulated digital wealth at exchanges and DAOs created enticing targets for other kinds of stress tests earlier.

The financial and market applications of the technology are already being observed (e.g., [11]), but for a voluntary, playful software to succeed and persist a lot more is needed. A stable-coin may have a backing organization behind it, but games live and die on player interest. Cryptogame companies need considerable numbers of players to create new, constant, emergent value through their gameplay, and thereby retain their users. The games have to be appealing in their content, playability, and business models. This means that the blockchain cannot be just an add-on. It has to produce actual added value to the player as well as to the developer company itself. The added value can be observed in the form of ingenious APIs that make gameplay and token-based content and ownership storage possible, but that is likely to be just the beginning. We regard it possible that for example the forthcoming, interesting options for security tokens are discovered for play, not for more traditional business forms, and then developed further by other types of companies.

One final topic, implicitly discussed throughout this chapter, is scalability (e.g., [29]). For a cryptocurrency, unless one is an idealist interested in changing the entire banking sector, partial adoption at times may be seen as sufficient (e.g., [30]). For a cryptogame to function in business terms, it needs both retention and scalability. So keeping an eye on the first blockchain games that actually reach stable player numbers above a few thousands will, in our opinion, be also a business advantage for blockchain developers. On the basis of these future success stories, there will likely be a rush of new research projects explaining the foundations and characteristics of such stories, in the same way we have seen studies emerging on other types of games. As cryptogames become more popular and their business models begin to be investigated in academic research more widely, we will learn whether their success was something created by the supposed added value of non-fungible tokens, or just solid, good game design.

As the points presented in this chapter arise partially from analysing cryptogames, and partially from critically analysing the marketing hype around them, we look forward to getting challenged in further research. This said, our belief in the value of the games together, as a kind of decentralized tool of their own, remains strong.

References

1. Marella, V., Upreti, B., Merikivi, J., Tuunainen, V.K.: Understanding the creation of trust in cryptocurrencies: the case of bitcoin. Electron. Mark. **30**(1), 259–271 (2020). https://doi.org/10.1007/s12525-019-00392-5
2. The Economist: cover on 31 October 2015. https://www.economist.com/weeklyedition/2015-10-31
3. Chohan, U.W.: Are cryptocurrencies truly trustless? In: Goutte, S., Guesmi, K., Saadi, S. (eds.) Cryptofinance and Mechanisms of Exchange: The Making of Virtual Currency, pp. 77–89. Springer, Cham (2019)
4. Zhao, J.L., Fan, S., Zheng, E.: Blockchain-enabled trust: the case of inter-firm dataflow (2018). AMCIS Presentation. https://aisel.aisnet.org/amcis2018/TREOsPDS/Presentations/71/
5. Nakamoto, S.: Bitcoin: a peer-to-peer electronic cash system. White paper (2008)
6. Morgan, J.P.: J.P. Morgan creates digital coin for payments (2019). https://www.jpmorgan.com/solutions/cib/news/digital-coin-payments
7. Peiró, N.N., Martinez, G.E.J.: Blockchain and land registration systems. Eur. Prop. Law J. **6**(3), 296–320 (2017). 10/ggzss9
8. Goodell, G., Aste, T.: A decentralized digital identity architecture. Front Blockchain **2** (2019). https://doi.org/10.3389/fbloc.2019.00017
9. Pfeiffer, A., Bugeja, M.: Introducing the concept of "digital-agent signatures": how self sovereign identity (SSI) can be expanded for the needs of industry 4.0. In: Dingli, A., Haddod, F., Kluever, C. (eds.) Artificial Intelligence in Industry 4.0. Springer, Cham (2020)
10. Kshetri, N.: Blockchain's roles in meeting key supply chain management objectives. Int. J. Inform. Manage. **39**, 80–89 (2018). https://doi.org/10.1016/j.ijinfomgt.2017.12.005
11. Stein, S.S.: Blockchain, Artificial Intelligence and Financial Services. Springer, Cham (2020)
12. Chohan, U.W.: The leisures of blockchains: exploratory analysis. SSRN J. (2017). https://doi.org/10.2139/ssrn.3084411
13. Scholten, O.J., Hughes, N.G.J., Deterding, S., Drachen, A., Walker, J.A., Zendle, D.I.: Ethereum crypto-games: mechanics, prevalence and gambling similarities. In: Proceedings of the Annual

Symposium on Computer-Human Interaction in Play CHI PLAY, pp. 379–389. Barcelona, Spain (2019). https://doi.org/10.1145/3311350.3347178
14. Gladyshev, V., Wu, Q.: Design for the decentralized world: democratization of blockchain-based software design. In: Marcus, A., Rosenzweig, E. (eds.) Design, User Experience, and Usability. Design for Contemporary Interactive Environments, pp. 74–86. Springer, Cham (2020)
15. Serada, A., Sihvonen, T., Harviainen, J.T.: CryptoKitties and the new ludic economy: how blockchain introduces value, ownership, and scarcity in digital gaming. Games Cult. **16**(4), 457–480 (2021). https://doi.org/10.1177/1555412019898305
16. Sky Mavis: Axie Infinity (2018)
17. Xaya: Taurion (2019)
18. Lehdonvirta, V., Castronova, E.: Virtual Economies: Design and Analysis. MIT Press, Cambridge (2014)
19. Xaya: Soccer Manager Elite (2020)
20. Harviainen, J.T., Ojasalo, J., Nanda Kumar, S.: Customer preferences in mobile game pricing: a service design based case study. Electron. Mark. **28**, 191–203 (2018). https://doi.org/10.1007/s12525-018-0285-6
21. Heimo, O., Harviainen, J.T., Kimppa, K., Mäkilä, T.: Virtual to virtuous money: video game business logic and virtue ethics. J. Bus. Ethics **153**(1), 95–103 (2018). https://doi.org/10.1007/s10551-016-3408-z
22. Martin, J.: Game review: axie infinity currently rules the pay-to-win-to-earn roost (2020). https://cointelegraph.com/news/game-review-axie-infinity-currently-rules-the-pay-to-win-to-earn-roost
23. Dapper Labs: Flow (2020)
24. Gerard, D.: Libra Shrugged: How Facebook Tried to Take Over the Money. United Kingdom (2020)
25. Terra Virtua: Terra Virtua (2020)
26. Blocklete Games: Blocklete Golf (2020)
27. Harviainen, J.T., Paavilainen, J., Koskinen, E.: Ayn Rand's objectivist ethics applied to video game business. J. Bus. Ethics **167**(4), 761–774 (2020). https://doi.org/10.1007/s10551-019-04159-y
28. Neopets: neopets (1999)
29. Jang, K.-S., Lee, O.: The design and development of a on chain game for scalability verification of blockchain platform. J. Digit. Converg. **18**(10), 253–263 (2020). https://doi.org/10.14400/JDC.2020.18.10.253
30. Hütten, M., Thiemann, M.: Moneys at the margins—from political experiment to cashless societies. In: Campbell-Verduyn, M. (ed.) Bitcoin and Beyond: The Challenges and Opportunities of Blockchains for Global Governance, pp. 25–47. Routledge, London (2018)

Fairness by Design: The Fair Game and the Fair Price on a Blockchain-Based Marketplace

Alesha Serada

Abstract It is often believed that blockchain technologies can ensure fairness in online transactions and interactions. What does 'fair game' mean in a blockchain-based game, which rules cannot be broken by design, and how does this relate to the concept of a 'fair price'? In this chapter, I use the example of the best known blockchain-based game CryptoKitties (2017) to explore the idea of a 'fair price' both in theory and in practice, and to connect it to the concept of fairness in games. I turn to the essential works on fairness and cheating in game studies and check whether game ethics is applicable to so-called 'money games' on blockchain. Theoretically, decentralization of blockchain technology supports the idea of fairness; however, in practice, the game follows the same grey moral code as the preceding online games and virtual worlds. I suggest that the applicable understanding of fairness can be found in the 'code is law' principle that underlines both normative game studies and the ideology of blockchain.

Keywords Blockchain · Blockchain games · Technology ethics · Fairness · Virtual economies

1 Introduction. Why CryptoKitties?

CryptoKitties [46] is one of the longest-running games that utilize blockchain technology. Initially, the game was built on the first version of the Ethereum platform [4], although it exists in a transitional state between different blockchains since the Ethereum network was clogged in 2020 [20]. According to the original game design, its players could trade game tokens—'kitties'—for the cryptocurrency Ether, and then exchange it for other cryptocurrencies, real world money and goods [10]. Unsurprisingly, the possibility to cash out earnings afforded ethically questionable behaviour such as speculation [26], in the form of seeking "the maximum benefits

A. Serada (✉)
University of Vaasa, Wolffintie 34, FI-65200 Vaasa, Finland
e-mail: aserada@uwasa.fi

A. Dingli et al. (eds.), *Disruptive Technologies in Media, Arts and Design*, Lecture Notes in Networks and Systems 382, https://doi.org/10.1007/978-3-030-93780-5_6

from market fluctuations" [26], while, potentially, manipulating the prices [43], often at the expense of less experienced participants. The game itself, however, attracted its share of devoted players; to distinguish between economic and playful activities in the game, Lee et al. suggest separating item-selling, which is most often driven by speculative motives, from item purchasing, gifting, and breeding, which constitute the entertaining communal aspect of the game [26].

Initially, the developers of the game pursued educational goals: namely, they aspired to make blockchain technologies accessible to general public [10, 36]. However, broad audiences initially became aware of CryptoKitties because of the ridiculously expensive purchases that happened in the first month of its existence [7]. Some suggested that some of these purchases could be connected to money laundering, pointing at the infamous "Cat named Dragon", which was sold for 600 ETH (US $170,000 at the time) [11]. Is it a fair price for this token? Is this kitty worth the money? In the remainder of this paper, I will explore the origins of the playful ethics that exist in blockchain-based games.

Not long after their introduction, digital 'kitties' entered less playful markets of crypto assets such as Uniswap [44]—I suggest this was the moment when the economic component of CryptoKitties eventually overshadowed it as a game. Still, as long as the game was consensual and financially profitable for at least some of its most economically-minded players, is it possible to call it unfair? After all, the authors who speculate about the possible future of blockchain suggest that this technology can enable algorithmic fairness and nurture pro-social behaviour [8, 24, 31, 32] due to decentralization and wider participation. Still, blockchain-based games remain a niche entertainment, not least, due to high participation costs that, in the case of most popular games, require several hundreds of US dollars to even start playing actively (e.g. [39]; this was also author's experience with CryptoKitties). It is true that blockchain technologies grant everyone equal access to 'crypto games'—which is considered fair in the design philosophy of blockchain [24]. However, as LaPointe and Fishbane describe in their ethical framework for blockchains, fairness of such design is not the same as equity of opportunities for different categories of users [24].

2 Artificial Scarcity on Blockchain

Can blockchain technology make a virtual marketplace fair? Generally speaking, unfair advantage can be gained through misinformation or concealment of crucial information about the trade by one side (which is often the case in CryptoKitties, see [34]). Blockchain platforms address this problem by offering transparency of all transactions across the blockchain (unless a specialized 'mixing' service is applied, which is effective but also costly). To achieve this, blockchain platforms utilize cryptographically protected immutable ledgers of all transactions kept on each node of the network and updated in a decentralized way (e.g. by reaching consensus between a particular set of nodes). To add another level of fairness, the Ethereum platform, which was launched in 2015, introduced so-called 'smart contracts' that automate transactions based on the pre-defined rule sets (some of such contracts can still be changed or terminated by the developers) [4]. Due to these features, blockchain is widely imagined as a technological enabler of trust [29] and even democracy [4] in the future.

In public blockchains, ethical behavior is further enforced by the proof-of-work architecture that incentivizes so-called 'miners' to verify only the rightful transactions. This architecture ensures trust between the parties that do not trust each otherwise—at least, this was the goal of Satoshi Nakamoto, the mythical creator of Bitcoin. Initially, Nakamoto designed a limited supply of Bitcoin, thus introducing the idea of value based on scarcity to the Bitcoin community; in his project, he hoped "to pick something that would make prices similar to existing currencies" [29]. This idea remains an integral part of a widely shared imaginary of Bitcoin as 'digital gold' [33], inspired by the cryptolibertarian agenda of the community, albeit not without emancipatory potential, as also seen from the left [1].

In short, the idea of scarcity-based value had been already established in cryptocurrencies, at least, on the semiotic level [28], long before cryptocurrency-based games. When the latter first appeared, their game designed incorporated 'artificial scarcity' as a seemingly natural basis for value, which can be seen from the white paper that described CryptoKitties' design. According to the creators of this game, the aim of their product was to explore digital scarcity and digital collectibles within the innovative space of blockchain technologies. An elaborate system of attributes and traits, 'genes', mutations and generations would ensure relative rarity, or, at least, highly uneven distribution, if not actual scarcity, of certain tokens in the game. However, the effective sales prices of tokens with different attributes were rarely aligned with their actual scarcity [37, 38]. Logically, if scarcity was the key to fair prices, it would be only reasonable to create a calculator of relative value based on scarcity. Nevertheless, despite several attempts in the community, this was never fully realized. The most commonly used calculator, KittyHelper, only shows 'the price floor', which the lowest price on the market for tokens with different traits and attributes. As of 2021, the average price for different categories of tokens is not present in its interface, because it appears to mean very little in the game. Same as on markets of real-life collectibles, big buyers are mostly concerned about the rarest and most valuable tokens, and such tokens are evaluated case by case, quite similarly to antiques or art works in the real world.

Rarity and uniqueness seem to be more productive, even if almost unpredictable, criteria to construct potential value of a 'cryptokitty' than scarcity. Uniqueness is literally in the name of a non-fungible Ethereum token (NFT), which each 'kitty' represents. Meanwhile, unpredictability is in the core logic of the game, which challenges luck, rather than skills, of its players. When the developers state that every blockchain token is unique, they suggest that there is always the chance that it will acquire higher value in the future—which is exactly the case of the Cat named Dragon. Paradoxically, it should be worthless according to the rules of the game, as it does not have any special attributes that would make it scarce. To the experienced player of the game, it is remarkable for its lack of any marketable traits that would allow it to sell it for any reasonable price. Still, the 600 Ether transaction is real and registered on blockchain with no possibility to revert it, be it a mistake. We may say that it is the immutable record of this trade that actually assigns the declared value to the otherwise worthless kitty. Since the historical event of its purchase, the Cat named Dragon has been holding the unique title of the most expensive 'kitty' in the world, and if its owner decides to resell it, there will be enough competing buyers to raise its price even higher.

Scarcity was not the only economical concept for the developers of the game to play around: they also introduced mechanisms to evaluate demand for particular tokens. While the open market was set up to determine the prices of 'second hand' tokens, the fair price of the 'first hand' tokens was decided by the reverse auction [10]. The 'smart contract' established the initial price of these so-called Generation 0 'kitties' based on the state of the market; the price slowly decreased with time, and players would buy the token when its price matched their assumption of its value. This would be fair enough to establish the price equilibrium, at least, on the first-hand market; however, as soon as the game was exposed to human players, its initial logic was immediately subverted in many speculative practices. Most economically-thinking players immediately took to 'flipping', which means reselling tokens for profit, typically short term [37]. In order to fetch the rarest or the cheapest kitties, most technically savvy players immediately started creating trading bots—a practice that appeared to be unfair to the players who could not afford a bot or did not know programming. This created a new form of information asymmetry that allowed most wealthy and educated players to win the game consistently in the economic terms [34]. However, when asked directly, these players still consider the game general fair.

Gradually, the community deciphered the complete 'genome' of CryptoKitties, built 'breeding calculators' and even 'autobreeders' based on it [12], which sometimes made human participation in the game rather inefficient. Still, all this was considered a meaningful part of the game—playful practices that were commonly accepted by the core player base as 'fair'. When I started my research of pricing in CryptoKitties, my goal was to find a meaningful distinction between 'fair' and 'unfair' price. It took me a long time to realize that my subjective ethics do not apply in the virtual world. For instance, many players set the prices far above the expected price on the market, and novice players sometimes buy into this—their loss! I first saw it as unfair, but it appears to be a universally accepted way to play—same as negotiating at a bazaar. When I asked the players about what they consider fair in the game, some of them suggested that an inflated price is not unfair, as this is also a part of play. Later, I asked the developer of another successful blockchain-based game in a private conversation, how they differentiate between honest players and speculators, and he replied: "All players are speculators! This is the essence of the game". The same can be said about CryptoKitties, which demonstrates once again that the concept of a 'fair price' appears to have a playful dimension that complicates any economic description of it, as well as ethical evaluation.

To sum it up, neither reverse auction nor calculation of relative scarcity would be sufficient to establish what is considered a 'fair price' in CryptoKitties. Same as at oriental bazaars [18], also used for video game trading [13], the pricing is always performative and relational. In the end, the idea of a 'fair price' seems to emerge from dynamic trading practices and social relations: generally, the community has some kind of a shared idea about what is fair in the game at each particular point of it. Although this question of 'right' and 'wrong' prices has always baffled the community of blockchain gamers, the inevitable information asymmetry and completely puzzling unpredictability of prices never 'spoilt the game' for its core players.

3 Why Are Games not Always Fair?

The question of (un-)fairness appears to be much more complicated in game studies, as well. A game is considered fair when everyone has an equal and fair chance to win the prize that is proportional to their input [5]. Cheating creates the unfair advantage for the cheater [9], although this unfairness is often perceived rather than calculated. In mainstream game studies, a fair game is often understood more broadly as an ethical game [40]. All its players not just get even or fairly proportional chances to win, they also have an equal right to enjoy the game in general. This right comes with the responsibility of being a 'virtuous player' who cooperates with other players, avoids cheating and confrontation other than in a rule-driven combat or competition. This is the desired norm in several foundational works of game studies [6, 21, 40], as well as many later developments of video game ethics.

Edward Castronova, a game economist, was one of the first to study fairness in virtual worlds. He developed his vision based on early multiplayer games such as Ultima Online and Second Life. Castronova is also one of the most consistently neoliberal scholars: he sees the purpose of play in accumulation of 'gaming capital' based on meritocracy. To comply with this purpose, virtual worlds must have pre-designed conditions and rules set by the 'coding authority'; these rules are deemed fair if the player accepts them. For instance, if a game world reproduces gender inequality, a player is free to leave this world for a different one with different rules "in which both genders are equally skilled and equally objectified", supposedly built by a nongovernmental organization to prove their point [6, p. 142]. Such ethical code is based on a set pre-defined external rules that do not account for internal conflicts of a social systems. Theoretical 'fairness' is easily distorted to sanction unfair and antisocial behaviour towards less privileged or simply less lucky members of the gaming community. The rules of such community are still negotiable, even if not always 'virtuous': for instance, the members of a particular online game to be unfair to female players, but punish cheating, deception and antisocial behaviour in other situations [41].

The question of practical game ethics has been reframed by Mia Consalvo in her empirical study Cheating. Gaining Advantage in Videogames [9]. Her studies of 'gaming capital' reveal that it does not always correspond to the rational ideal of 'meritocracy', but is rather a construct within the existing economic and social relations that arise in and around gaming cultures. Gaining and especially maintaining and acknowledging symbolic capital in games implies that the gamer shall not cheat.

Gaming communities usually have some kind of a shared vision of fairness, as well as communal agreements that support pro-social interactions and enable joyous and relatively conflict-free collective play.

Let us take a closer look at some of the multiple understandings of cheating that Consalvo's research reveals. For instance, in the eyes of some players, unfairness comes from gaining an advantage from the external information that does not belong to the space where the game challenge is taking place. From this viewpoint, cheating is use of any other sources of information apart from the affordances of the game itself [9]. Meanwhile, if we apply this understanding to the game of CryptoKitties, we will find it unfair by design. It seems to have run on informational asymmetries from the start: for example, unequal opportunities between the players who can and cannot code, and, of course, the privileged position of those who already had a history of owning cryptocurrencies. Finally, it would be impossible to play the game without referring to external sources of information in the first year of its existence: the game mechanics were obscure, and it did not even have a complete player guide [12]. Players had to leave the 'magic circle' of the game and ask the community how various features of CryptoKitties worked. Even though there was very little drama in the community as compared to an average server of a multiplayer game, there was always the risk that the early adopters would play a trick or two on the newcomers, which would typically cost them certain amounts of the cryptocurrency Ether. In the end, limiting the access to the external sources of information would make it almost impossible to play the game, as its very point was exploration of the novel blockchain technology.

Unsurprisingly, the inherent informational asymmetry between seasoned players and newcomers have not resulted in a particularly healthy market. Due to it, the prices on the marketplace were often intentionally inflated far above the generally agreed benchmark with the hope to catch a clueless newbie or an ignorant 'crypto whale', or generally to manipulate the price level for certain tokens. All this is nothing new in multiplayer games: Consalvo mentions similar fraudulent transactions, aimed to inflate prices of in-game objects, in the early digital children's game Whyville in 2006 [9, p. 117]. Allegedly, the same techniques has been consequently used on the emerging markets of NFTs in 2021.

Yet another conceptualization of fairness, according to Consalvo, states that the game is fair when it is played truthfully to the game code and design. Consalvo connects this idea to the 'code is law' principle, which she finds in Lawrence Lessig [9, p. 90]. This is where we discover the genetic link between the understanding of 'fairness' in video games and on cryptocurrency markets. Many blockchain adopters reproduce the same belief as Lessig, which originated from the early cyber-anarchism [16]. To them, 'code writers' of the internet are also its 'lawmakers'. To 'play fair' is to act according to the affordances of the code, allegedly designed in an 'algorithmically fair' way. This may seem redundant, as the rules set in smart contracts cannot be broken by design [4]; however, interests of other players or market participants are not a part of this rule-based technocentric ethical system. According to its adopters, exploitation of weaknesses of other players is fair as long as the pre-established rules are followed and the original code of the game is not corrupted. It must be noted, though, that communities of blockchain adopters have their own ethical code that prohibits openly anti-social behaviour such as stealing another person's cryptocurrency or 'mining' it on someone else's property without permission [3, p. 96]. Speaking of CryptoKitties, there have always been ambiguous cases that tested even this, rather straightforward 'law is code' principle, such as trading bots or exploits that allowed gaining Ether by force-executing the 'birthing contract'. However, there are no coded rules in the game that would discourage speculation. The code of the game puts no limits on the price one might want to value their property for sale, so they are free to name any price (such as, 600 Ether for the Cat named Dragon), and other players are free to pay the price if they have the money. It is the responsibility of players to reduce information asymmetry and 'do their own research' before conducting a trade.

4 Designing Fair Competition

According to the normative game ethics, players become moral beings by following the rules of the game. This remains true if the rules of the game violate personal ethical code of players in real life, e.g. in violent games. From the perspective, as presented by Miguel Sicart, the player-subject is defined by the rules of the game, and her main virtue is fidelity: "The fidelity of the player is present as long as her actions are coherent with the game rules and the game world, and do not contradict a rule" [40, p. 74]. When we study games as ludic systems that consist of rules, we have to accept that in-game rules override extraludic ethics—which, by the way, can also be used in productive and positive ways to create ethically interesting games, as Sicart later describes in more detail.

Following the rules is particularly important in multiplayer games. Sicart stresses that cheating breaks the game experience: it is detrimental not only for the cheater, but for other players as well. By prioritising the rule system of the game, its ethics appeal to the broader technocratic "code is law" principle. Such rule-based game ethics rightfully dominate in highly competitive games: most of such games are not about reaching the consensus between the participants and certainly not about the situation when everybody wins. To the contrary, competitive games always include an element of justifiable suffering, such as the risk to lose, to feel frustrated and humiliated. The ethical threshold for the potential harm is decided by the community, and, just as any social rules, it is always highly contextual and constantly negotiated. Nguyen and Zagal come up with two criteria for ethical competition: "It depends, first, on consent and second, on the motivational set-up of the players" [30], namely, their willingness to endure a certain degree of violence. As a result, there is no ethical problem in spoiling the game for the disadvantaged (i.e. less experienced) members in many multiplayer games, as long as everybody follows the rules of the game itself.

This leaves designers of multiplayer games with the almost impossible challenge to make the competition as fair as possible, at least, at the level of the game system, which some players will inevitable try to exploit, while others will complain and threaten the developers. Skewed chances in the game are the usual source of discontent (the second one being game developers, personal and professional qualities). As early as in "Synthetic Worlds" [6], Castronova observes early gaming communities trying to negotiate with developers about a fairer and more balanced game—a scene familiar to anyone who has ever participated in any multiplayer game, from Destiny 2 to FarmVille 2. Mia Consalvo provides a similar description of the world of Final Fantasy XI. Speaking of virtual economies, players may even require developers to prevent other players from gaining an unfair advantage by financial means [22].

Is fair competition even possible? It appears that Roger Caillois disputed the possibility of fair competition in games long before digital games even came into being [5]. As long as players of a game originate from different social stratas, they will always have advantages and disadvantages predefined by their access to wealth, education and training, before the game has even begun. The only case of absolutely fair competition appears to be gambling, and exactly this mechanics lies at the core of CryptoKitties [35]. Gambling, however, is unethical when it becomes an addiction, and this side of blockchain-based games also needs urgent research.

In summary, the rule-based approach to game ethics, which prioritises games as systems, is not without its merit: it provides novel creative opportunities and relative simplicity of designing and running games. Still, we might wonder how following external rules makes us a 'virtuous player': these rules could have been set by a potentially immoral subject or a corporate entity such as a business firm, who produce almost all popular games. Besides, this type of game ethics does not protect the underprivileged, such as the players who do not have enough symbolic or financial capital to participate in a game to the fullest degree—or just those who have entered the game at a later stage. In the end, CryptoKitties was not designed as a competitive game—it was envisioned as creative exploration of blockchain technologies that everyone could try for themselves. Unfortunately, this Utopian project ended up in much frustration for casual and not particularly wealthy players.

5 Second Morality?

Literature on cheating suggests that buying in-game wealth and power on external markets for real money is often considered an unfair advantage [9, 22]. Interestingly, 'crypto game' developers are very eager to give the players the right to buy and sell their in-game rewards for real money, which contrasts with the traditional ethics of early MMO games, where buying and selling items violated implicit and even explicit rules of the game[9, p. 164]. Today the ability to buy and sell in-game upgrades, power-ups and particularly mighty weapons became the basis for extensive monetization [27], especially in the free-to-play games that are sometimes criticized as 'pay to win'. Upon closer inspection, almost all blockchain-based games are 'pay to win' by design, but this is the topic for another time.

As for now, almost all online games have internal virtual economies and markets for various digital commodities. Prices are usually set by developers, publishers, or game marketing specialists—in other words, some kind of a ‘coding authority’. Some multiplayer games (for instance, Team Fortress 2 (2007) and Counter Strike: Global Offensive (2012) published by Valve) and virtual worlds, such as Second Life (2003), have peer-to-peer markets where players can trade in-game objects and set their own prices. As soon as the prices get out of control of the ‘coding authority’, speculation with digital commodities flourishes, sometimes despite all technological limitations and preventive measures, on which honest and virtuous players insist. An early case of such speculation in massive multiplayer online games has been first thoroughly described in an autoethnographic study by Julian Dibbell, later published as a personal narrative [14].

For the most part, the spirit of CryptoKitties is reminiscent of these ‘grey’ markets of virtual goods sprawling around the Steam game platform and even external ‘black’ markets of game items and characters in multiplayer games such as World of Warcraft. These game markets of the not quite forgotten past are notorious for cheating, speculation and legal disputes [2, 15, 17, 19, 42]. On the other hand, libertarian economists, such as Castronova and his like-minded colleagues Lastowka and Hunter, tend to conceptualize in-game speculation in mostly positive economic terms [25], with the hope that the equilibrium of prices is achievable, and virtual economies will eventually mature into the state of efficient self-regulation. This is also the goal of blockchain-based games, although the most long-living one seems to demonstrate stagnation rather than maturation in economic terms.

Interestingly, the same game may be found ethical or unethical when seen from either a normative (rule-based) or a descriptive (player-centric) approach in game studies. From the cyberlibertarian perspective, Second Life is an ethical virtual world, because its rules allow players to gain wealth proportionally to their time and input [6]. However, such claims should not be taken by face value, as many similar virtual economies, such as EVE Online, also allow players to gain wealth much faster by piracy and grey market trade. To Sicart, EVE Online is an ethical game because everybody has equal chances in it. He is aware of piracy, but he suggests that the informal pirate code is at least as important as the official rules of the game (refusing to be a pirate in Eve Online may disrupt gaming experience of other players). Learning to be a cosmic pirate is such an important part of the game that players lose their ‘virtual subjectivity’ if they do not engage into it to the degree required by the community [40, p. 72]. This is in line with the idea of different subjectivities that an individual wears ‘inside’ and ‘outside’ of virtual worlds.

An interesting question here is at which point the ‘inside subjectivity’ becomes responsible for their deeds according to the jurisdiction that their physical body belongs to. This is inevitable when the players who have lost their virtual property seek justice at a real life court [15, 22]—and it becomes even clearer when real-life financial crimes are conducted by the use of the affordances of virtual worlds. It has been noted many times that design of Second Life affords financial crimes in the real world [42]. The same, to a much greater degree, is true for the affordances of blockchain technologies [22, 42], which calls for more research in its highly speculative markets.

In this light, the final point to consider is the double morality of game owners and publishers. Initially, developers of CryptoKitties criticized the practice of manipulating scarcity on virtual markets as unfair [10] (see also [27] on its negative impact on virtual economies). Blockchain technology provided an antidote: the code of CryptoKitties should prevent even the developers themselves from creating new 'scarcities'. However, Dapper Lab started doing exactly that later the same year. Since 2018, developers have been regularly introducing new limited edition 'fancies' to keep the game alive. These new categories of tokens have always had higher value than regular 'kitties' on the inflated market. This is very different from projects like CryptoPunks [23], which have a limited amount of tokens that cannot be exceeded, and these tokens are in fact only getting more expensive. Their actual scarcity has led to a curious project that creates a digital double for an existing CryptoPunk on Ethereum, so another layer of value is developed on the crypto market, which offers financial rewards in the form of bounties, lending and depositing particularly rare tokens that may not even change hands in the process [45], also somewhat similarly to the real world art market. While developers of CryptoKitties pointed at CryptoPunks as an example of a fizzled-out project in 2018, the latter are still highly valued in 2021, and they still deliver the initial agenda of their creators, as any piece of 'crypto art' should aspire to do.

6 Conclusion. Ethics of Blockchain Versus Ethics of Its Adopters

Can fairness be reinforced by immutable ledgers, 'smart contracts' and financial incentives, as the latest projects of blockchain-based governance suggest? A 'fair price' on a crypto market does not have a fixed transparent value secured in blockchain, and is acquired in a process akin to bargaining on an oriental market [18]. In this paper, I hope to have demonstrated that such price is established in a playful collective action, as a part of supposedly 'fair game'. This playful attitude may be characteristic to blockchain-based markets in general.

In technocentric communities of 'crypto gamers', game ethics is mostly derived from the same 'code is law' principle that is so important for the blockchain community. In game studies, the 'code is law' principle corresponds to the rule-based (normative) perspective that places game rules above real life ethics and laws. Ideally, a fair game encourages all its participants to play in an honest and rational way, and this encouragement is not limited to game rewards: it includes, for example, the sense of belonging to a certain community and acquiring symbolic capital. In practice, when game ethics is externalized as a system of rules set in code in an 'unreal' virtual world, nothing—not even one's moral compass—can prevent the player from cheating and abusing others, as soon as there is space for free play in the codified game system. This is also noticeable in CryptoKitties. In theory, the game was designed in accordance with the principles of 'ethical games' [40]: blockchain is expected to ensure transparency of transactions, and its 'smart contracts' are set up to produce artificial scarcity. In practice, we can observe ubiquitous speculation, cheating and other deviations from prescribed prosocial behaviour. Still, this does not spoil the game for the most faithful players.

Long before blockchain-based games, empirical studies have consistently revealed the anti-social side of multiplayer game worlds. In fact, the in-game morality of its players does not seem to differ much from their general ethical disposition [17]; cheating and speculation inevitably emerge in collective play and flourish in some communities that tolerate or even encourage such behaviour as a part of the game. Virtual worlds apply 'double morality' to the actions of in-game and out-of-game subjects, which may afford offensive behaviour in the game and even financial crimes in the real world. Still, researchers believe that game communities can, and will, rule themselves in a democratic way to prevent cheating and abuse. This is one of the contributions that blockchain-based games can possibly offer to more generalized game research. Almost every blockchain-based game in 2018–2019 demonstrated very active collective decision making among players and developers, even though, probably, it was due to the community rather than the technology. I suggest that this can be a potentially valuable tendency to follow in game development—as long as reliance on community does not become a tool to extract free labour from its members, which would be a different kind of unfairness.

In gaming, blockchain is suggested to provide 'trustlessness' by removing intermediaries, such as game publishers, from interactions between players. However, it does not prevent neither players nor developers from exploiting each other's trust and creating new information asymmetries. In this regard, blockchain-based games do not differ from early online multiplayer games. The ethics of blockchain-based games is not defined its 'smart contracts' and immutable ledgers. Instead, it is the collective responsibility of their developers and users. There is nothing special about blockchain that could prevent its unethical application—in fact, some of its properties, such as decentralization and anonymity, actively invite cheating, speculation and gambling. In the end, game publishers are responsible for their product regardless of technology they use, and they have to comply with legal regulations when playful unfairness bleeds into the real world. Not all players are 'virtuous', neither all of them want to be, and the same is true for game developers. If the rules of a 'fair game' allow speculation, then any price is 'fair', but this also means that, much like in gambling, the only safe strategy to avoid losses is to avoid the game altogether.

Acknowledgements This chapter was written with support from the Nissi Foundation for doctoral students involved in commercial studies.

References

1. Allon, F.: Money after blockchain: gold, decentralised politics and the new libertarianism. Aust. Fem. Stud. **33**(96), 223–243 (2018). https://doi.org/10.1080/08164649.2018.1517245
2. Blitz, M.J.: A first amendment for second life: what virtual worlds mean for the law of video games. Vanderbilt J. Entertain. Technol. Law **11**(4), 779–821 (2009)

3. Bohr, J., Bashir, M.: Who uses bitcoin? An exploration of the bitcoin community. In: 2014 Twelfth Annual International Conference on Privacy, Security and Trust, pp. 94–101. IEEE, Toronto, ON (2014). https://doi.org/10.1109/PST.2014.6890928. http://ieeexplore.ieee.org/document/6890928/
4. Buterin, V.: Ethereum White Paper. A Next Generation Smart Contract & Decentralized Application Platform (2013). https://ethereum.org/en/whitepaper/
5. Caillois, R.: Man, Play, and Games. University of Illinois Press (1961). Google-Books-ID: mQfIAAAACAAJ
6. Castronova, E.: Synthetic Worlds: The Business and Culture of Online Games. University of Chicago Press, Chicago (2005)
7. Chohan, U.W.: The leisures of blockchains: exploratory analysis. SSRN Electron. J. (2017). https://doi.org/10.2139/ssrn.3084411
8. Ciriello, R.F., Beck, R., Thatcher, J.: The Paradoxical Effects of Blockchain Technology on Social Networking Practices. San Francisco, CA (2018)
9. Consalvo, M.: Cheating: Gaining Advantage in Videogames. The MIT Press (2009)
10. CryptoKitties: CryptoKitties: Collectible and Breedable Cats Empowered by Blockchain Technology. White Pa-Purr (2018)
11. CryptoKitties: herding one-million cats (2018). https://www.cryptokitties.co/blog/post/herding-one-million-cats. Section: Blockchain
12. CryptoKitties: learn how our last mystery smart contract works (2019). https://www.cryptokitties.co/blog/post/learn-how-our-last-mystery-smart-contract-works. Section: Blockchain
13. Deka, M.: Bazaars and video games in India. BioScope: South Asian Screen Stud. **7**(2), 172–188 (2016). https://doi.org/10.1177/0974927616668005
14. Dibbell, J.: Play Money: Or, How I Quit My Day Job and Made Millions Trading Virtual Loot. Basic Books, New York (2006)
15. Fairfield, J.A.T.: Anti-social contracts: the contractual governance of virtual worlds. McGill Law J. **53**, 427–475 (2008)
16. Filippi, P.D., Hassan, S.: Blockchain technology as a regulatory technology: from code is law to law is code. First Monday (2016). https://doi.org/10.5210/fm.v21i12.7113. https://firstmonday.org/ojs/index.php/fm/article/view/7113
17. Gabriels, K., Bauwens, J., Verstrynge, K.: Second life, second morality? In: Virtual Worlds and Metaverse Platforms: New Communication and Identity Paradigms pp. 306–320 (2012). https://doi.org/10.4018/978-1-60960-854-5.ch020. https://www.igi-global.com/chapter/second-life-second-morality/55415
18. Geertz, C.: The bazaar economy: information and search in peasant marketing. Am. Econ. Rev. **68**(2), 28–32 (1978). https://www.jstor.org/stable/1816656
19. Heeks, R.: Understanding "gold farming" and real-money trading as the intersection of real and virtual economies. J. Virtual Worlds Res. **2**(4) (2009). https://doi.org/10.4101/jvwr.v2i4.868. https://journals.tdl.org/jvwr/index.php/jvwr/article/view/868
20. Jordan, J.: Daily activity down 87%: is tether killing ethereum gaming? (2020). https://dappradar.com/blog/daily-activity-down-87-is-tether-killing-ethereum-gaming. Library Catalog: dappradar.com Section: Games
21. Juul, J.: Half-Real: Video Games Between Real Rules and Fictional Worlds. MIT Press, Cambridge, MA (2005). OCLC: 260129192
22. Kim, T.: The predecessors of bitcoin and their implications for the prospect of virtual currencies. PLoS One **10**(4),(2015). https://doi.org/10.1371/journal.pone.0123071
23. Labs, L.: CryptoPunks (2017). https://www.larvalabs.com/cryptopunks
24. Lapointe, C., Fishbane, L.: The blockchain ethical design framework. Innov. Technol. Gov. Glob. **12**(3–4), 50–71 (2019). https://doi.org/10.1162/inov_a_00275
25. Lastowka, G., Hunter, D.: The Laws of the Virtual Worlds. SSRN Scholarly Paper ID 402860. Social Science Research Network, Rochester, NY (2003). https://papers.ssrn.com/abstract=402860

26. Lee, J., Yoo, B., Jang, M.: Is a blockchain-based game a game for fun, or is it a tool for speculation? An empirical analysis of player behavior in crypokitties. In: The Ecosystem of e-Business: Technologies, Stakeholders, and Connections. Lecture Notes in Business Information Processing, vol. 357, pp. 141–148. Springer (2019)
27. Lehdonvirta, V., Castronova, E.: Virtual Economies: Design and Analysis. MIT Press (2014). https://www.jstor.org/stable/j.ctt9qf5t6
28. Maurer, B., Nelms, T.C., Swartz, L.: "When perhaps the real problem is money itself!": the practical materiality of bitcoin. Soc. Semiot. **23**(2), 261–277 (2013). https://doi.org/10.1080/10350330.2013.777594. http://www.tandfonline.com/doi/abs/10.1080/10350330.2013.777594
29. Nakamoto, S.: Bitcoin: A Peer-to-Peer Electronic Cash System (2008)
30. Nguyen, C.T., Zagal, J.P.: Good violence, bad violence: the ethics of competition in multiplayer games. In: Proceedings of 1st International Joint Conference of DiGRA and FDG. Digital Games Research Association, Dundee (2016)
31. Piasecki, P.J.: Gaming self-contained provably fair smart contract casinos. Ledger **1**, 99–110 (2016). https://doi.org/10.5195/ledger.2016.29. http://ledgerjournal.org/ojs/index.php/ledger/article/view/29
32. Pitt, J., Diaconescu, A.: The algorithmic governance of common-pool resources. In: From Bitcoin to Burning Man and Beyond: The Quest for Identity and Autonomy in a Digital Society, pp. 130–142. ID3, Off the Commons Books, Amherst (2014). OCLC: 1034977162
33. Popper, N.: Digital Gold: Bitcoin and the Inside Story of the Misfits and Millionaires Trying to Reinvent Money, p. 00008. Harper, New York (2015)
34. Sako, K., Matsuo, S., Meier, S.: Fairness in ERC token markets: a case study of cryptoKitties. arXiv:2102.03721 [cs] (2021). http://arxiv.org/abs/2102.03721
35. Scholten, O.J., Hughes, N.G.J., Deterding, S., Drachen, A., Walker, J.A., Zendle, D.: Ethereum crypto-games: mechanics, prevalence, and gambling similarities. In: Proceedings of the Annual Symposium on Computer-Human Interaction in Play, pp. 379–389. ACM, Barcelona (2019). https://doi.org/10.1145/3311350.3347178. http://dl.acm.org/doi/10.1145/3311350.3347178
36. Serada, A.: Cryptomarkets gamified: what can we learn by playing cryptoKitties? In: DiGRA Proceedings. Digital Games Research Association DiGRA (2020)
37. Serada, A.: Vintage cryptokitties and the quest for authenticity. In: Proceedings of IEEE CoG, Copenhagen (2021)
38. Serada, A.: Why is cryptoKitties (not) gambling? In: International Conference on the Foundations of Digital Games (FDG '20), Bugibba, Malta, 15–18 Sept 2020. https://doi.org/10.1145/3402942.3402985
39. Servando, K., Sayson, I.C.: This video game is turning the pandemic jobless into crypto traders. Bloomberg.com (2021). https://www.bloomberg.com/news/articles/2021-08-25/axie-infinity-how-game-is-turning-pandemic-jobless-into-crypto-nft-traders
40. Sicart, M.: The Ethics of Computer Games. MIT Press, Cambridge, MA (2009). OCLC: ocn248539580
41. Ståhl, M., Rusk, F.: Player customization, competence and team discourse: exploring player identity (co)construction in counter-strike: global offensive. Game Stud. **20**(4) (2020). http://gamestudies.org/2004/articles/stahl_rusk
42. Stokes, R.: Virtual money laundering: the case of Bitcoin and the Linden dollar. Inf. Commun. Technol. Law **21**(3), 221–236 (2012). https://doi.org/10.1080/13600834.2012.744225
43. Szado, E.: Defining speculation: the first step toward a rational dialogue. J. Altern. Invest. **14**(1), 75–82 (2011). https://doi.org/10.3905/jai.2011.14.1.075
44. Uniswap: Uniswap (2018). https://uniswap.org/
45. Wrapped Punks (2020). https://wrappedpunks.com/
46. Zen, A.: CryptoKitties (2017). https://www.cryptokitties.co/

Horizontal Scalability of Blockchain Games Using the GSP Model

Daniel Kraft

Abstract In recent years, blockchain has emerged as an interesting technology not just for digital currency but also more complex decentralised applications including computer games. Smart-contract platforms like Ethereum are a popular choice for developers of blockchain games. However, due to their architecture they are unsuitable for complex on-chain computations, as those would lead to unfeasibly high gas costs. Consequently, most blockchain games today are either trivial in terms of game mechanics or run on a centralised server, defeating the original purpose of using a blockchain. In this paper, we present an alternate architecture based on game-state processors (GSPs). This model decouples computations from the core blockchain layer (miners and full nodes), effectively sharding computation and storage of state between the different applications and games. With this, a blockchain platform can be scaled horizontally to support many and highly-complex games with fully decentralised, on-chain computations.

Keywords Blockchain gaming · Decentralised gaming · Play-to-earn · Game-state processor · Scalability

1 Introduction

Only a few years after Nakamoto's invention of blockchain technology [21] in 2008 his idea was being applied to things beyond just a currency. The first such application was Namecoin [14], but already soon after that a fully decentralised computer game based on blockchain was described [8] and launched with Huntercoin [10]. Apart from trivial gambling (e.g. SatoshiDice [4]), this was the first blockchain game ever

D. Kraft (✉)
Xaya Team, Autonomous Worlds Ltd, Zurich, Switzerland
e-mail: daniel@xaya.io

A. Dingli et al. (eds.), *Disruptive Technologies in Media, Arts and Design*, Lecture Notes in Networks and Systems 382, https://doi.org/10.1007/978-3-030-93780-5_7

created, and it also pioneered the concept of *human mining*. Soon after, a second decentralised game (Motocoin [30]) introduced another interesting concept, *proof-of-play*. Based on the original ethos and cypherpunk spirit of the Bitcoin community at the time, those early projects were indeed fully decentralised like Bitcoin itself.

Later on, smart-contract platforms like Ethereum [1] made it much easier to create blockchain-based games. During the Bitcoin bullrun of 2017, CryptoKitties was such a smart-contract game that gained widespread attention [27], and made larger audiences aware of blockchain as a tool for computer games. Unfortunately, the monolithic design of smart-contract platforms makes them unsuitable for decentralised and complex game logic—this means that most blockchain games to date are either almost trivial or using the blockchain mainly to store game assets, while the game itself still runs on centralised infrastructure similar to traditional, non-blockchain online games. CryptoKitties itself arguably falls in both categories at the same time; the main interaction with the game is breeding digital cats, which can hardly be seen as a particularly deep or complex game concept. But even with this simple design, a crucial part of the game (namely the pictures associated to cats, i.e. the phenotype derived from the on-chain genes) is not decentralised and instead hosted on the game studio's servers. Enjin is even an entire platform built around the concept of centralised games that use blockchain-based assets [26]. Even for centralised games, the use of blockchain-based assets in the game certainly has many benefits. For instance, it makes it much easier for users to hold, transfer and trade their assets, even if the game developer was not able or willing to build a dedicated in-game marketplace. For Ethereum-based assets, OpenSea [11] has developed into a de-facto standard, universal marketplace where such assets can be traded.

However, game assets always have to be considered in the context of the games themselves. Assets of a centralised game may be owned by the users in a trustless manner on a public, decentralised blockchain (like Ethereum); but if the developers of the game decide to shut down the game servers, ban a user or inflate the supply of a supposedly rare item, then items will become worthless even if the game developers may not be able to take them directly off a user. Especially with smart contracts, it is nowadays quite common for the developers to give themselves special power to upgrade the contract or otherwise control the system at will; such backdoors have in the past been used to shut down supposedly decentralised applications such as the IOTA network [7] or the tBTC application [12].

Therefore, it is important to distinguish between *true ownership* of assets in fully decentralised games and just *non-custodial ownership* of assets in centralised games. The latter is only an evolution, which makes asset-ownership and trading easier; the former is a true revolution, which shows the real disruptive potential of blockchain technology. This can be compared to how Bitcoin enabled the emergence of value without any trusted, central instance for the first time in history, while "permissioned" blockchains are just a handy tool to optimise some processes [13]. For a more in-depth discussion of the benefits of fully decentralised gaming, see [28], specifically the sections about CryptoKitties and hybrid architectures.

Xaya is a blockchain, platform and open-source project that strives to enable fully decentralised, on-chain games that are still able to feature highly complex game logic and computations. In contrast to games built solely on smart-contract platforms, this is enabled by decoupling the core blockchain from the potentially heavy computations and states of individual games. This is possible through an architecture based on *game-state processors* (GSPs), which we will describe in this paper.

Note that while we work on this topic ourselves to build decentralised computer games, the term *game* in the following can just as well be understood to mean any type of decentralised application (dApp), not necessarily restricted to entertainment purposes.

2 Scalability for Blockchain Games

At least since the blocksize debate in Bitcoin [5], it is clear that scalability in general is a hot topic and potential issue for any blockchain-based system. In fact, blockchain as a data structure is very inefficient in terms of scalability; only because it enables otherwise-impossible full decentralisation should it be used at all [31].

For "simple" applications of blockchain technology like Bitcoin itself, the main scalability issue is the data that needs to be put into blocks, transmitted through the network, and stored forever in the blockchain. But for more complex applications (e.g. smart contracts on Ethereum) and especially non-trivial games, another important bottleneck is the computing power needed to execute the contract or game logic. In many cases of complex games, this may be the limiting factor rather than the volume of raw data. Therefore, it makes sense to distinguish between two slightly different types of scalability in this context.

2.1 Vertical Scalability

Even a monolithic design (where a single game runs on its own blockchain) can easily run into scalability limits. For instance, Fig. 1 shows the size of the Huntercoin blockchain over the first 18 months after its launch. Especially right at the beginning, the growth (i.e. transaction volume) was tremendous, considering also that this was back in 2014 where even Bitcoin itself was not yet hitting the blocksize limit. It became clear that this growth was not sustainable, and a couple of network-wide changes were implemented to address the issue (some of which are marked with vertical lines in the chart). As can be seen, they took effect and slowed down blockchain growth. However, this solution is far from perfect, as it came with restrictions and severe changes to how Huntercoin could be used and played. Also, since these changes were tailored specifically to Huntercoin, they cannot be applied directly to other games.

Huntercoin was created as an experiment to see how blockchains can be used for MMO-type games. As it became clear that some scalability solution is needed, we developed *game channels* [19] as an off-chain scaling technology. Inspired by payment channels [23] and developed independently from state channels [20] at around the same time, they allow complex game interactions (and not just payments) to be done off-chain while keeping the whole system fully decentralised and trustless. This enables massive scalability *within a single game* as well as *near-realtime blockchain-gaming experiences*. In the following, however, we want to focus on a different aspect of scalability, and refer to [19] for more details about game channels and vertical scalability.

2.2 Horizontal Scalability

If a blockchain platform is not specific to a single game and instead built to run many games and applications in parallel, then the limits imposed on each individual application have to be even tighter. For instance, Ethereum as the leading smart-contract platform uses *gas fees* to charge each user according to the computations their transactions require in smart contracts. As a consequence, it becomes prohibitively expensive to execute anything beyond the most trivial game logic. For instance, it is even impossible to directly check whether or not a position in chess is check mate

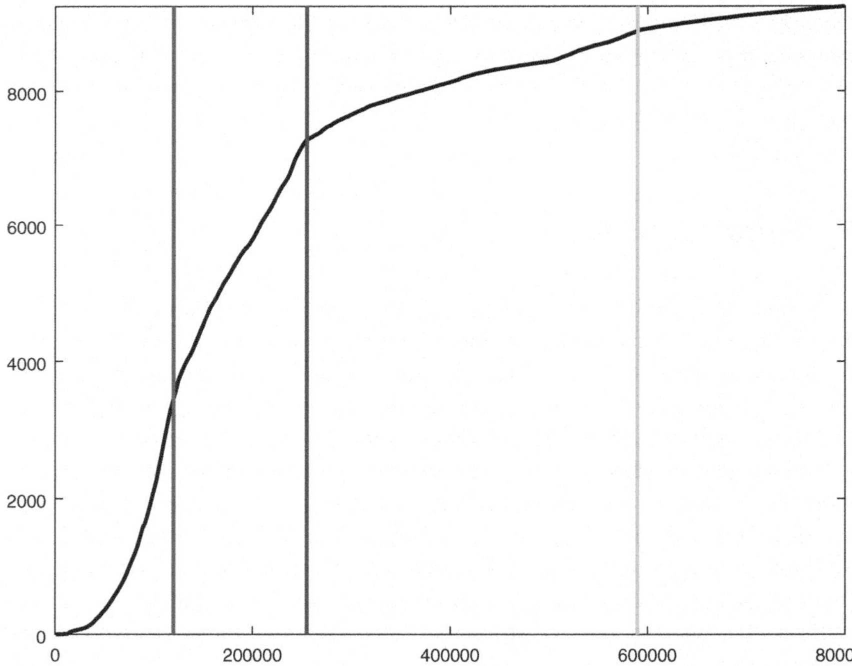

Fig. 1 Size of the Huntercoin blockchain in MiB during the first 800,000 blocks (about 18 months) after launch. The vertical lines indicate changes to the network rules

[15]; and while it is certainly challenging for a human to play well in chess, the rules themselves are much simpler than e.g. the computations required for a real MMO strategy game. Similarly, CryptoKitties famously clogged the Ethereum network due to its massive traffic after launch [16].

Strictly speaking, both of these are of course examples of single games running into scalability limits, and thus could be considered vertical scalability issues. However, a decentralised network is certainly able to verify check-mate positions. But if the network as a whole must be able to execute the computations of many other smart contracts as well, then the limits have to be a lot more restrictive, which is exactly what happens with Ethereum.

This is especially the case if the cost in computation (and perhaps state storage) outweighs the raw transaction size stored in the blockchain, as is the case for the chess example and many other potential gaming applications. For these situations, the monolithic approach of smart-contract platforms is not very cost effective or even outright prohibitive. To enable *horizontal scalability* of a blockchain to *many and complex applications*, a different model is required, which we will discuss below.

3 The Role of Miners

It is a common misconception, especially in mainstream media, that "the job of a Bitcoin miner is to validate transactions, and in exchange get a portion of the reward in BTC for doing it" [24]. Of course, Bitcoin miners typically also validate transactions before they put them into blocks; if they did not do this, they would risk losing money for mining invalid blocks. (In fact, in 2015 an incident related to "SPV mining" actually caused miners who did not properly validate blocks to lose block rewards in this way [25].) However, from the network's point of view, *all full nodes* (which are typically a superset of the miners) validate transactions.

The real function of mining is to achieve consensus with proof-of-work (PoW). More specifically, through mining the network establishes a *time order* in which transactions occurred, thus enabling to determine which of a possible set of conflicting transactions is the "correct" one, solving the *double-spending problem*. This is described as a "timestamp server" already in the original Bitcoin whitepaper [21].

4 Game-State Processors

Now all necessary preliminaries are in place, and we can discuss how the GSP architecture allows to decouple the potentially expensive computations for some game from the core blockchain network, and thus greatly improve the network's horizontal scalability.

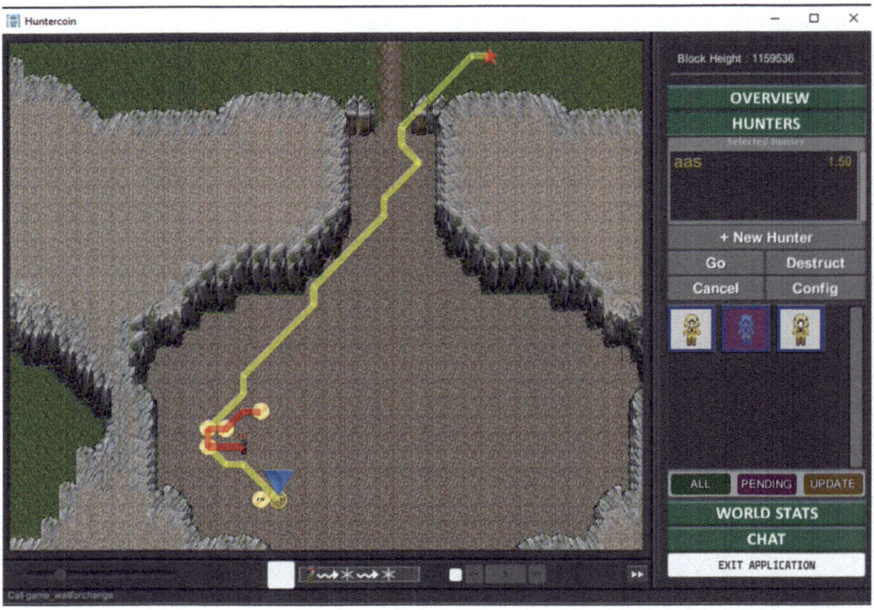

Fig. 2 Screenshot of Huntercoin, showing some of the game's basic concepts

4.1 Game States

Blockchains in general can be interpreted as a *state-transition system* [1]. Associated to each block is a current state. It can be computed deterministically from the state associated to the previous block, the transactions inside the block as "update data", and the network's hardcoded rules (the state-transition function). In the case of Bitcoin, the state is the current ledger, represented by the UTXO set. The updates applied are transactions that spend some inputs (remove them from the UTXO set) and create new outputs (that get added to the UTXO set). Other blockchains typically have more complex states and state-transition functions, e.g. Namecoin or Ethereum.

Figure 2 shows a screenshot of Huntercoin. In the game, players can move around with their avatars ("hunters") on a map. They are able to pick up coins placed on the map, and then need to bring them to a banking location while others may try to attack them and steal the coins. Once banked, the coins appear in the user's wallet, and act like an ordinary cryptocurrency. Consequently, the *game state* of Huntercoin contains like Bitcoin's the ledger of banked coins in form of an UTXO set; however, it also contains the state of the game world, i.e. the positions of all hunters on the map, their waypoints and all coins that are not yet banked. This state can be affected by each player by sending their moves as transactions in the blockchain. These moves are the creation of new hunters, changes to their waypoints (thus controlling movement) and an "attack" action.

It is important to understand that the state itself is never stored in the blockchain, only moves (player actions) are. The state is computed independently from those moves by each network node for themselves. Since mining ensures consensus about the set and order of moves that have occurred and each node is using the same game rules (state-transition function), everyone is guaranteed to obtain the same state as well.

4.2 Decoupling States from the Core Blockchain

Let us now briefly summarise what we have learned so far: It is not necessary to store the actual game state on chain; instead, it is enough to store the *moves* on the blockchain, and then have every node compute the game state independently. The only thing that is required to make this work and ensure everyone arrives at the same state is to ensure that they are using the same state-transition function (i.e. are using the same consensus rules and follow the same hard forks), and that PoW mining yields a network consensus on the sequence of moves that have taken place. But on the other hand, it is *not* necessary for miners to actually interpret what those moves mean in the context of a particular game, and thus it is also not necessary for miners to know any current game state (or even the rules of individual games).

This insight allows us to *decouple* computation of game states from the underlying core network; thus, we can split the architecture of such a platform into two layers:

1. The core layer consists of the P2P network of nodes, the actual blockchain, and PoW mining of transactions. It also needs a cryptocurrency to pay for transaction fees and in this way protect against blockchain spam (by having a fee market). This layer allows arbitrary data for moves to be put on chain, without validating or interpreting them in any game-specific way. In the case of Xaya, the core layer also manages a concept of "user accounts" (based on Namecoin names), and ensures through digital signatures that only the owner of an account can send moves for it.
2. Each game running on the platform provides a *game-state processor* (GSP): This is a software process that runs alongside the core blockchain node, reads the sequence of moves from the blockchain, and uses the game-specific state-transition function to compute the current state. This can then e.g. be used to show the state to the user in a game frontend, or for other purposes (perhaps as part of an exchange that allows trading of assets from one of the games).

The crucial observation here is that miners and full nodes are only required to handle the first part of this. All heavy computation for a particular game (as well as storing that game's state) is done by the GSP instead. And the GSP for some game only needs to be run by users who need to know the state of that game, i.e. who are actively interested in the game (e.g. playing it). This allows a single blockchain platform to support hundreds or thousands of complex games in parallel, because each user is only required to do computations for the games they are using themselves.

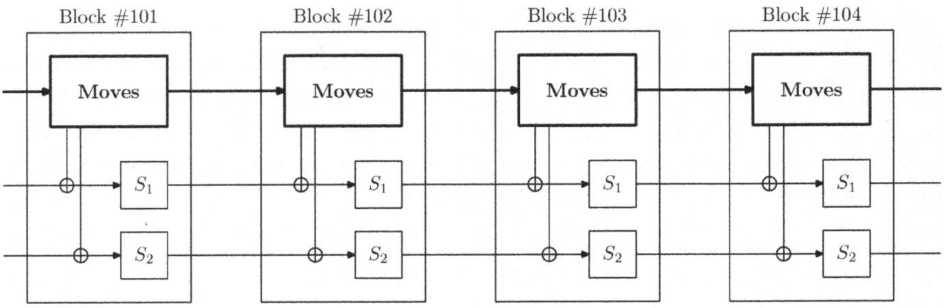

Fig. 3 Schematic diagram of two different games (GSPs) running on top of a single core blockchain (bold)

In some sense, this can be seen as (partial) *sharding*, namely sharding of computation power and storage (but not of the raw blockchain data).

This model is in complete contrast to a monolithic platform like Ethereum, where every full node (and thus miner) always has to perform all computations of the entire network. Hence, a platform based on GSPs is highly scalable horizontally, while a monolithic platform is not.

This split into a core layer and multiple GSPs on top of it is sketched in Fig. 3: The core blockchain corresponds to the bold parts in the diagram, i.e. blocks that contain just some moves. Those moves are then used by two GSPs running on the platform to derive the states S_1 and S_2 associated to each of the blocks. The $\oplus$ indicates the state-transition function (game logic) for each game, which derives the new state from the previous state and the current block's moves.

Note that each GSP is purely local, and does not concern the core network at all. This has an interesting consequence: Even if at some point in time no one is playing a particular game at all, the game still carries on (it is unstoppable), although there does not exist any instance of the current state at that time. If at a later time someone starts playing again and runs the GSP, they will compute the game state for the entire history (including those blocks when no GSPs were running) and "re-materialise" the game world. This also means that unlike platforms that have a "one blockchain per game" model (e.g. Planetarium [6]), there is no need to incentivise and worry about nodes running GSPs all the time.

Let us also remark that the decoupling of states from a core blockchain (which is used to achieve consensus) is not a completely new idea. A couple of overlay protocols have been built on top of Bitcoin in the past, e.g. Mastercoin [29] (now called Omni) or Counterparty [22]. To the best of our knowledge, however, all of these platforms were built to place a specific use case on top of Bitcoin and take advantage of the Bitcoin network's security. In our proposal, we use the overlay architecture for a different purpose, namely for horizontal scaling of a custom blockchain. In a sense, this can be compared rather to *rollups*, which have become the scaling solution of choice in Ethereum [3]: Each GSP can be thought of as its own rollup, except radically simplified and not relying on any complex security assumptions and mechanisms to bridge state back to the main chain.

5 Drawbacks of Game-State Decoupling

We have seen that decoupling the game state from the core network leads to drastic improvements in scalability. On the other hand, it also means that the core network does not and can not (by design) fully verify game-specific operations, which imposes two main restrictions on games built with this architecture.

5.1 SPV Security

SPV (simplified payment verification; see [21]) is a method to run cryptocurrency wallets without the need to download and verify the entire blockchain. It is used by popular tools such as Electrum or mobile Bitcoin wallets. SPV wallets assume that all transactions inside the longest (most work) chain of blocks are valid without verifying them for themselves. This works as long as the majority of mining power is honest, since miners have a strong incentive to only include valid transactions into blocks (as discussed previously).

Strictly speaking, this of course remains true for GSP-based blockchains. However, for them SPV security obviously only applies to things validated by the core network (e.g. CHI transactions and name operations in the case of Xaya). Since the game-specific interpretation of move data is left to GSPs, the mere presence of a particular move in the longest chain does not automatically imply that the move has a particular effect on the game state or is even valid for a specific game.

For example, Alice could send a move that indicates "send a Vorpal sword to Bob", and it could be confirmed and included into the longest chain of blocks just fine. However, it may still be invalid with respect to the game, e.g. because Alice does not own such a sword. Hence, Bob must not rely on the sword being transferred to him just because Alice's transaction has been confirmed. Instead, he has to check the game state himself by running a GSP.

However, it is still possible to build mobile and light clients for GSP-based games (just not with the SPV technique alone). Depending on the desired balance between decentralisation and convenience, users may simply trust a remote GSP that is run by the game developer, especially if they do not yet have a lot of money at stake inside the game. If their interest and stake in the game increases, they can at any later time decide to run their own GSP for more security and reliability instead. The game developer (or a third party) can even provide ready-made VM or container images for GSPs, so that also non-technical users are easily able to run their own GSP on a cheap VPS and still play on mobile devices.

It is even possible to design a system that allows for a balance between decentralisation and convenience that is strictly between the two extremes of "trusting a single remote GSP" and "running your own GSP". For instance, users could query multiple remote GSPs and only accept a current game state if all of them agree. Or the operators of remote GSPs could sign the states they deliver, so that a wrong answer can be proven to the network; with this, it is possible to build a "GSP-as-a-service" system based on deposits and penalties similar to some of the more recent proof-of-stake consensus algorithms [2] or optimistic rollups like Arbitrum [17].

5.2 No Control Over the Coin

The second restriction implied by state decoupling is that GSPs and specific games cannot control the native on-chain currency directly; this is in direct contrast to smart contracts on monolithic platforms (e.g. contracts on Ethereum have their own ETH balance). As coin transactions are part of the core network layer and thus need to be validated by all nodes, it is not possible that any coin transfers are conditional on the state of some game (which is not known by those nodes). It is, however, possible for a GSP to *observe* coin transactions that happen on the core layer and then update the game state accordingly (for instance, mint in-game assets whenever coins are paid to the game developer's address). This is a direct consequence of the unidirectional flow of information between the two layers (from the core network to GSPs, but not the other way round).

This restricts how value can flow into and out of games through the blockchain's native currency (especially for human-mining/play-to-earn games). It is, however, possible to add a game-specific currency (e.g. gold pieces, credits, gems, ...) to the game state. Together with all other game assets, balances of this currency can then be fully controlled by arbitrary rules coded into the GSP. As long as this game currency can be exchanged to some fiat or other cryptocurrency, it will enable free flow of value again.

The latter, finally, is easily possible: Centralised exchanges can integrate an in-game currency just as easily as a "normal" cryptocurrency; all they have to do is run the game's GSP. Additionally, the core blockchain can be designed to facilitate atomic trades between its coin and game assets, so that trustless and decentralised exchanges can be built as well. In the case of Xaya, *atomic name updates* [18] can be used for this purpose; they are based on *atomic name trades* originally proposed for Namecoin [9].

6 Conclusion

We have seen that scalability is a major concern for blockchain projects in general, but even more so for blockchain-based games. Monolithic approaches based on smart contracts are popular for blockchain games today, but they lead to games that are either centralised (with only assets on the blockchain) or have trivial rules and shallow game play. In other words, Ethereum may not be the best choice for building a fully decentralised game.

Hence, we proposed to decouple game-logic processing from the core blockchain network. This GSP architecture improves horizontal scalability drastically and, in particular, allows to build complex and still fully decentralised games. Game-state decoupling comes not without its own trade-offs, though, as it means that the core network is not aware of game-specific states and rules. This leads to some restrictions, but we believe that they can be dealt with very well in practice using, among other techniques, atomic transactions for trading between the native coin of the core blockchain and in-game assets.

With the methods described, it is possible to bring blockchain games to the next level of their evolution, and build highly complex but still fully trustless, scalable and decentralised games and applications.

Acknowledgements The author is employed by Autonomous Worlds Ltd, the startup backing the Xaya project. This work was created through that employment.

References

1. Buterin, V.: A Next-Generation Smart Contract and Decentralized Application Platform (2013). https://ethereum.org/whitepaper/
2. Buterin, V.: A Proof of Stake Design Philosophy (2016). https://medium.com/@VitalikButerin/a-proof-of-stake-design-philosophy-506585978d51
3. Buterin, V.: A rollup-centric Ethereum roadmap (2020). https://ethereum-magicians.org/t/a-rollup-centric-ethereum-roadmap/4698
4. Buterin, V.: The bitcoin gambling diaspora. Bitcoin Magazine (2013). https://bitcoinmagazine.com/articles/the-bitcoin-gambling-diaspora-1375548799
5. Caffyn, G.: What is the bitcoin block size debate and why does it matter? CoinDesk (2015). https://www.coindesk.com/what-is-the-bitcoin-block-size-debate-and-why-does-it-matter
6. Chung, I.: Every game this South Korean startup makes has its own blockchain. CoinDesk (2019). https://www.coindesk.com/every-game-this-south-korean-startup-makes-has-its-own-blockchain
7. Cimpanu, C.: IOTA cryptocurrency shuts down entire network after wallet hack. ZDNet (2020). https://www.zdnet.com/article/iota-cryptocurrency-shuts-down-entire-network-after-wallet-hack/
8. Colosimo, A.: Chrono Kings—a game/world within the block chain (2013). https://bitcointalk.org/index.php?topic=262599.0
9. Colosimo, A.: Fair exchange of a Namecoin name to NMC (2013). https://forum.namecoin.org/viewtopic.php?t=1316
10. Colosimo, A.: Huntercoin—Worlds First Decentralized Game/World on the Blockchain (2014). https://bitcointalk.org/index.php?topic=435170.0
11. Dale, B.: 'Ebay for cryptoKitties' raises $2 million from all-star VCs. CoinDesk (2018). https://www.coindesk.com/ebay-cryptokitties-raises-2-million-star-vcs
12. Dale, B.: Bug forces shutdown of bitcoin-backed ethereum token tBTC. CoinDesk (2020). https://www.coindesk.com/bug-forces-shutdown-of-bitcoin-backed-ethereum-token-tbtc

13. Dale, B.: Decentralized or nothing: Song duels IBM over blockchain hype at SXSW. CoinDesk (2019). https://www.coindesk.com/decentralized-or-nothing-song-slams-ibm-over-blockchain-hype-at-sxsw
14. Durham, V.: Namecoin—a distributed naming system based on bitcoin (2011). https://bitcointalk.org/index.php?topic=6017.0
15. Grau, P., et al.: Lessons learned from making a chess game for ethereum (2016). https://medium.com/@graycoding/lessons-learned-from-making-a-chess-game-for-ethereum-6917c01178b6
16. Hertig, A.: Loveable digital kittens are clogging ethereum's blockchain. CoinDesk (2017). https://www.coindesk.com/loveable-digital-kittens-clogging-ethereums-blockchain
17. Kalodner, H., Goldfeder, S., Chen, X., Weinberg, S.M., Felten, E.W.: Arbitrum: scalable, private smart contracts. In: 27th USENIX Security Symposium (USENIX Security 18), pp. 1353–1370. USENIX Association (2018). https://www.usenix.org/conference/usenixsecurity18/presentation/kalodner
18. Kraft, D.: Atomic trading (2019). https://github.com/xaya/xaya/blob/master/doc/xaya/trading.md
19. Kraft, D.: Game channels for trustless off-chain interactions in decentralized virtual worlds. Ledger **1**, 84–98. https://doi.org/10.5195/ledger.2016.15
20. McCone, R.: Ethereum lightning network and beyond (2015). http://www.arcturnus.com/ethereum-lightning-network-and-beyond/. Accessible in the WayBack Machine
21. Nakamoto, S.: Bitcoin: A Peer-to-Peer Electronic Cash System (2008). https://bitcoin.org/bitcoin.pdf
22. PhantomPhreak: Counterparty—Pioneering Peer-to-Peer Finance—Official Thread (2014). https://bitcointalk.org/index.php?topic=395761.0
23. Rapidly-adjusted (micro)payments to a pre-determined party. https://en.bitcoin.it/wiki/Contract#Example_7:_Rapidly-adjusted_.28micro.29payments_to_a_pre-determined_party
24. Raza, A.: A new bitcoin bug could cause chaos in the crypto world. Invezz (2020). https://invezz.com/news/2020/06/01/new-bitcoin-bug-found/
25. Rizzo, P.: Double spending risk remains after July 4th bitcoin fork. CoinDesk (2015). https://www.coindesk.com/double-spending-risk-bitcoin-network-fork
26. Sinclair, S.: Enjin launches game development platform on ethereum. CoinDesk (2020). https://www.coindesk.com/new-enjin-platform-aims-to-make-it-easier-for-game-developers-to-integrate-blockchain-assets
27. Takahashi, D.: CryptoKitties explained: why players have bred over a million blockchain felines. VentureBeat (2018). https://venturebeat.com/2018/10/06/cryptokitties-explained-why-players-have-bred-over-a-million-blockchain-felines/
28. Thomson, E.: The thick and thin of blockchain gaming architectures (2019). https://medium.com/@edward.thomson/the-thick-and-thin-of-blockhain-gaming-architectures-a51795156420
29. Willet, J.R.: MasterCoin: New Protocol Layer Starting from "The Exodus Address" (2013). https://bitcointalk.org/index.php?topic=265488.0
30. WilliamLie2: Motocoin: the first proof-of-play cryptocurrency (2014). https://bitcointalk.org/index.php?topic=591724.0
31. Woo, W.: 'Bitcoin is the future, blockchain is bullshit': Andreas Antonopoulos preempts consensus. Bitcoinist (2018). https://bitcoinist.com/blockchain-bullshit-andreas-antonopoulos/

AI

A Criticism of the Technological Singularity

Alexander K. Seewald

Abstract The compelling story of self-improving machines which become exponentially smarter up to inconceivable intelligence has inspired some of the best science fiction literature [1–3], but is also taken seriously by many researchers. This story is however based on empirical observations of seemingly exponential processes such as Moore's law in the semiconductor industry, and contains multiple fallacies concerning self-improvement of intelligent systems (including humans), which upon close look are implausible. Deep Learning has been heralded as a major step in this direction, however a closer look again shows many open issues with this approach, leading us to conclude that we deciphered only a small part of one method which nervous systems *may* use to create intelligent behaviour; that seemingly simple tasks like image classification and segmentation are still AI-complete; and that true Artificial General Intelligence (AGI) still lies at least several centuries in the future. But even an AGI would not be able to exponentially self-improve without further advances. These fallacies abound in science fiction literature as well as in scientific papers, and we will illustrate our analysis with appropriate examples.

Keywords Singularity · Criticism · Artificial General Intelligence · Moore's Law · Cybernetics · AI History

1 Introduction

The technological singularity, or Intelligence Explosion Hypothesis (IEH), is a theoretical concept where exponential improvements in intelligence (e.g. a factor of 2 every year) yield corresponding speedups of the underlying system where these speedups were obtained, making the next exponential improvement step by the same system exponentially faster (e.g. a factor of 2 every 6 months, every 3 months, every

A. K. Seewald (✉)
Seewald Solutions, Lärchenstraße 1, 4616 Weißkirchen a.d., Traun, Austria
e-mail: alex@seewald.at
URL: https://www.seewald.at

A. Dingli et al. (eds.), *Disruptive Technologies in Media, Arts and Design*, Lecture Notes in Networks and Systems 382, https://doi.org/10.1007/978-3-030-93780-5_8

1.5 months, ...), thus yielding exponential intelligence improvement steps in exponentially shorter time, rapidly reaching a point where the total intelligence of such a system reaches a mathematical singularity (e.g. $\frac{1}{0}$). Self-recursive improvement is thus an essential feature of this concept. The effect is in each case accelerated technological change, leading to a rapid transition to a state where the human condition would be challenged, and the future would be unpredictable.

The concept is not new. Already in the 1950, John von Neumann [4] claimed that accelerating technological progress was approaching a mathematical singularity beyond which human affairs would radically change.[1] Good [5] postulated an intelligence explosion by recursive self-improvement of a machine intelligence, which he called an *ultraintelligent* machine. Lem [6] published a science fiction novel about a military AI that recursively increases its intelligence repeatedly, rapidly approaching just such a singularity. Two years later, Vinge [7] contended that a rapidly self-improving AI would quickly approach a technological singularity beyond which reality would be unpredictable, and later elaborated on this in [8], where he predicted the singularity to happen between 2005 and 2030. Solomonff [9] described six AI milestones he believed would lead to AI approaching practically infinite intelligence in finite time, assuming again a self-recursive improvement (in the form of: having computers makes it cheaper to create newer, more powerful computers) and a constant expenditure on exponentially decreasing hardware costs. Moravec [10] predicted that in 2038—due to Moore's law and similar trends—robotic reasoning and behavior would exceed that of humans. In 1990, examining evidence outside robotics, Kurzweil [11, 12] came to similar conclusions and put the singularity at around 2045. Yudkowsky [13] differentiates three different singularity schools:

1. *Accelerating Change*: Technological change follows (often exponential or super-exponential) smooth curves, so the arrival of new technologies—or when they cross key thresholds (such as Artificial General Intelligence)—can be predicted with reasonable precision.
2. *Event Horizon*: To predict a superintelligence's plans, you need to be at least as smart yourself. Thus the future after the creation of superintelligence is completely unpredictable. This is most similar to the definition by Vinge in [8].
3. *Intelligence Explosion*: The positive self-improvement feedback cycle of intelligence does not stop but triggers further and ever faster intelligence improvements of similar magnitude and creates a superintelligence before it hits physical limits.

He notes that their claims contradict one another: for example, if we use Moore's law to predict computing performance in 2099, we contradict both *Event Horizon* (which state that we cannot know the future after superintelligences) and *Intelligence Explosion* (because progress will run faster once smarter-than-human minds and nanotechnology are integrated into the process). However these comprehensive discussions do not at all indicate that any of these schools describe even remotely plausible scenarios.

[1] This was at a time of excessive optimism in AI research, similar to the present (2021).

Sandberg [14] compares models of the technological singularity and concludes that mathematical models for growth exhibit at least exponential growth as this is the signature of linear self-coupling terms. He adds that if efficiences of scale exist, superexponential growth or finite time singularities appear to be generic. He notes that mathematical singularities are likely indicators for transitions to other domains of growth, or that unmodeled factors will become relevant close to the point and are not meant to be taken at face value. He also notes that the most solid finding is that even small increasing returns in a growth model can produce radical growth, and that if mental capital (embodied in humans, artificial intelligence or posthumans) becomes relatively cheaply copyable, extremely rapid growth is likely to follow. Concluding, he notes that there is a lack of models of how an intelligence explosion could occur and that available evidence show that human experts are usually weak at long-term forecasting even without apparent singularities.

There are two distinct ways such a singularity is presumed to happen:

1. *Hard takeoff*: Runaway intelligence, recursive speeding-up self-improvement *in the blink of an eye*.
2. *Soft takeoff*: An exponential speedup similar what we have experienced due to Moore's law.

As we will see, a hard takeoff is extremely unlikely given the very strict requirements; however even a soft takeoff over a reasonable timeframe of 10–20 years also seems rather unlikely. A sufficiently slow soft takeoff taking several centuries would be undistinguishable from the status quo and we will therefore not address it.

There are many issues with past work on this topic, so we will first only mention a few of them: For example, above work mostly assumes that increases in research funding—one working example of a self-improving system—yield exponential increases in research progress while these are actually sublinear [15]. This is also supported by the finding of Hanson [16] who found that past progress in research is not perceived as increasing over time by the researchers themselves. In fact, based on the much more reliable past rate of progress estimated by researchers themselves, Hanson [17] (pp. 61–63) estimates at least two to four centuries until an Artificial General Intelligence—i.e. a human-level AI—becomes available.

It is sometimes claimed (e.g. in [8]) that intelligence amplification in humans could yield comparable results. However it seems obvious that the speedups necessary for a runaway intelligence explosion are not applicable to biological nervous system since the necessary self-improvement loop is not feasible for biological entities, at least at ever decreasing temporal lengths. Therefore we do not address this topic here.

Lastly, it should be noted that from an evolutionary standpoint, our facility for rational human thought has not evolved for drawing valid and statistically accurate conclusions, but rather to impress and persuade other humans by telling compelling stories [18]. So it does not come as a surprise that the singularity—a story of self-improving machines with unlimited potential—has a wide range of proponents as well as detractors. Please bear this in mind for the following discussion.

2 Singularity Requirements

There are three key requirements for any kind of technological singularity.

1. Recursive self-improvement must be feasible, regardless of the intelligence level to start with. It is not at all clear that recursive self-improvement is actually possible. If we go from the most intelligent system known so far—humans—cognitive introspection is a bad indicator of how things are actually implemented in our brains, and can only barely inform improvements in intelligence. Systematic experiments were and are still needed to determine good ways to improve mathematical abilities, memory, and logic even temporarily [19]. General intelligence is practically not improvable.[2] In machine learning after decades of research, there seem to be no true examples of unsupervised self-learning systems that automatically improve without explicit or implicit external feedback. In each case where machines have seemingly improved themselves (e.g. speech recognition, faster CPUs and GPUs, image classification and segmentation, self-driving cars, spam filtering, ...), this process was enabled by human ingenuity at exponentially increasing scales of effort and cost.
2. Recursive self-improvement must be feasible at a cost at most linear in the achieved improvement. For example, to make a system twice as smart, we must incur at most a linear cost $O(x)$ with x corresponding to the initial smartness of the system. So making a system x times smarter must be only x times as hard.[3] Otherwise the maximum achievable intelligence grows much more slowly, or even converges asymptotically to a limit never to be exceeded—surely the opposite of a singularity. However indications abound that it is actually much harder to make a system smarter, on the order of $O(x^2)$ or even $O(e^x)$. For example, increasing research funding yields much less than linear increases in research progress [15], indicating that costs are much more than linear w.r.t progress. In fact vulnerabilities such as Meltdown and Spectre [22] clearly indicate that previous costs at least in computer chip design have been underestimated in that they did not account for very subtle significant errors in previous chip designs.
3. At a linear cost of recursive self-improvement, we would still have to have an additional exponential process to achieve the necessary speedup for a singularity with *hard take-off*. For the *soft take-off* scenario the previous two requirements are sufficient. For example, we could exponentially increase the available budget for such a project, or a system twice as intelligent could only need half the cost to improve by another factor of two.[4] Another usual candidate is the exponentially increasing number of transistors in microchips at shrinking costs due to Moore's law. However this relies on (1) that is it possible to parallelize the recursive self-improvement process to an arbitrary number of subprocesses which can all

[2] Stumpf [20] finds correlations of 0.89 to 0.96 for adults over a seven-year period.

[3] [21] has produced a nice graph to visualize this.

[4] I.e. exponentially sped-up self-improvement – to some extent a strong version of the second requirement.

run in parallel, (2) for the exponential process to hold at least until the earliest timepoint where the singularity can happen. Both assumptions are doubtful: (1) ignores all communication costs between processors, but these also increase superlinear with the number of processors[5]; and (2) relies heavily on Moore's law which will hold only for another one to two decades at most (see Sect. 3).

The most critical requirement is the second one: Achieving the singularity assumes a sublinear effort for intelligence improvement. E.g. to achieve double intelligence, a constant effort is needed regardless of the previous intelligence level. Even a slight exponent on the effort will prevent a singularity from happening. Naam [23] has argued against such a superexponential runoff. He has pointed out that we already see recursive self-improvement by superintelligences, such as corporations like Intel which utilize the collective cognitive resources of tens of thousands of humans as well as millions of older CPUs to design the next CPU generation, and have been doing this for decades. However, this has not led to a hard takeoff; rather, it has led to a very soft takeoff in the form of Moore's law. Naam [21] further points out that the computational complexity to create higher intelligence may be much greater than linear.

It may be argued that evolution on earth has been running just such an experiment at recursive self-improvement for the last 3.5 billion years and has been relatively successful, even creating several human-level intelligences from scratch, all but one of which have already died out.[6] From this we can infer that recursive self-improvement is possible over sufficiently long time spans at the level of species—not necessarily at individual level. In fact the process used by evolution, *Survival of the Fittest* [25], is not reliant on individual intelligence at all. Taking a close look at the many organisms (*solutions*) it is clear that most of them are in an evolutionary dead-end and further optimization is likely to be increasingly difficult (see e.g. [26], pp. 104–105).

But although the biosphere contains approximately 16 million different species and $5 * 10^{30}$ living cells and has been "running" for about 3.5 billion years, optimizing all species in parallel, human-level intelligence appeared only relatively recently around 6 million years ago (i.e. after 99.83% of elapsed "runtime"). From this we may infer that creating the higher levels of intelligence so essential for the singularity is exceedingly hard, and generating them from scratch may be very costly.

It may also be argued that evolution does not optimize for intelligence as it may not be a survival trait.[7] However the one long-standing result compatible with this view—that highly intelligent people have fewer children than less intelligent ones—has been reevaluated by Parker [28] who found no such difference after reanalyzing

[5] Unless we connect all pairs of processors together, which then quickly runs into problems of available space as the number of connections is quadratic in the number of processors.

[6] Unfortunately the extinct homonids seem to have been the more peaceful ones, e.g. Kwang Hyun [24] states in his conclusion that the genetic basis for aggressiveness and hyperactivity originate with humans, and human interbreeding with Neanderthals led to more peaceful behavior in their hybrid offspring. This is likely the reason why they are extinct.

[7] Very entertaining argued for in [27]—but please consider last paragraph of Sect. 1 on the human weakness for compelling stories.

earlier work, and who speculated that sibling density—leading to less individual educational attention and less monetary and educational resources—might also explain these differences. Again we note that intelligence cannot be *improved* over the level obtained by nature and early nurture influences and is remarkably stable in adults, however many effects (nutritional, emotional, education, ...) can *stunt* intelligence below the level that could be obtained under optimal conditions, so these results do not contradict the ones mentioned earlier on intelligence being practically not improvable [20].

It is also sometimes claimed that the number of different ways to the singularity (such as Artificial Intelligence, Intelligence Amplification, hive minds, and so on) increases the likelihood of it to happen. However, this is a rather simplistic view. Notably, if only one of above requirements cannot be met, the singularity will never happen.

Concluding, we find all three mentioned conditions to be implausible. Recursive self-improvement is empirically only shown at species or population level, and may not be feasible on the level of individual intelligences at all, as human intelligence—the only known working example—is practically not improvable. There are also indications from evolutionary biology that recursive-self improvement is very costly. Finally, the additional exponential process needed to actually get to arbitrarily high intelligence levels—Moore's law is very often used here—merits a closer look as it is a central point of quite a few singularity proponents.

3 Moore's Law Revisited

Moore's law has often been cited and used in projections as a support for the impending technological singularity. Even when it is not directly referred to, Grace [29] notes that software costs roughly track hardware costs in many domains, so estimates on such progress—even when not directly referencing Moore's law—still indirectly rely on its exponential improvements. Tuomi [30] makes a very good case that the increasing number of transistors on chips confounds two different patterns; we added a third somewhat obvious observation that may not be common knowledge.

1. In 1965, initially Moore's law—then a empirically grounded hypothesis on the future of semiconductor production technology—referenced minimum-cost chips, i.e. those that could be most efficiently produced. However, ten years later in 1975 it was changed to reference maximum-complexity chips, i.e. at the limit of the production technology, which changes both timing (higher values will tend to be reported earlier) and the transistor count itself (which will also be higher, perhaps beyond profitability).[8]
2. From about 1998, the chip industry started to put cache memory on its chips to improve performance. This increases the transistor count dramatically with only

[8] Also, maximum-complexity chips sometimes had bad price/performance ratio and were hard to sell, occasionally never coming to market at all.

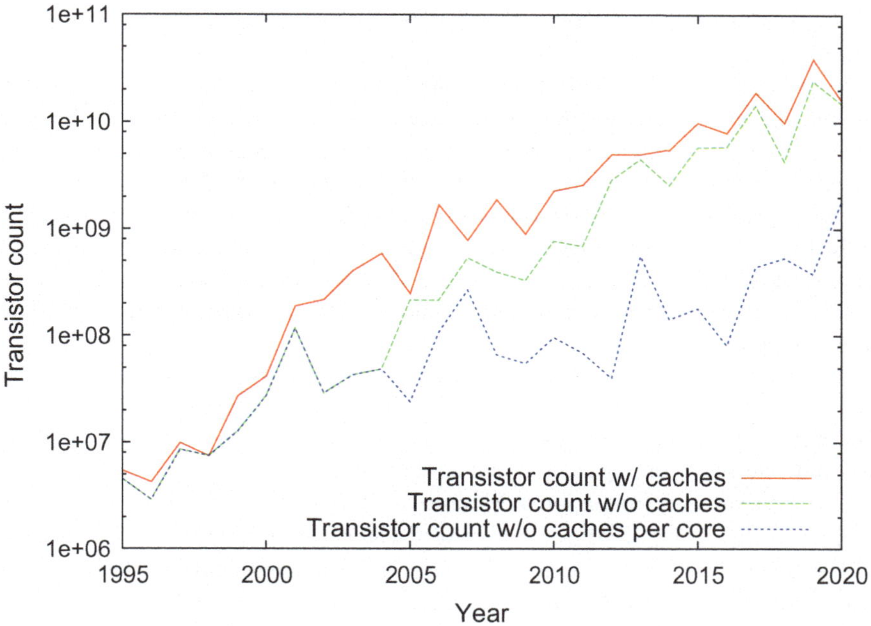

Fig. 1 Transistor counts between 1995 and 2020

modest increases in design complexity, as cache memory units are relatively uniform and simple compared to other CPU units. This explanation also applies to some extent to graphical processing units (GPUs) used extensively in Deep Learning. For example, the last DEC Alpha chip had 90% of the transistors corresponding to cache memory [31] and the Intel DualCore Itanium-2 from 2006 had 96.28% cache memory [32].

3. From about 2005, semiconductor manufacturers started to fill available die area with multiple copies of chip units (*cores*) that are essentially identical, thus increasing transistor count with only small increases in design complexity.

We cannot easily verify (1) but it was only at the beginning of Moore's law and is no longer relevant. However, we can with some effort confirm (2) and (3) by closely analyzing the processors with the highest transistor counts in each year from 1995 to 2020 (data from [31]). We manually determined transistor count, cache size and number of cores for each processor. To reduce effort, we only analyzed one processor per year, namely the one with the highest transistor count. Figure 1 shows uncorrected transistor count, transistor count when removing approximate cache transistors[9] (compensating for effect 2), and the transistor count per core (compen-

[9] We estimated seven transistors per bit, which was current in 1999 but may be down to four transistors at present. However on an exponential scale a factor of two corresponds to a constant shift and does not effect the rest of this analysis. Also our method ignores cache logic and thus slightly underestimates the number of cache transistors, which can also be seen from [32] (entry

sating for effects 2 and 3). As can be seen, cache contributed a large proportion of transistors between 2000 and 2005, and its contribution has been shrinking since then. The number of transistors by core shows only a very weak exponential growth and may even become asymptotic in the near future. The doubling time for by-core transistor counts is already 3.84 years, almost twice the doubling time for uncorrected Moore's law (of two years). The doubling time for 1995–2007 in our period is 2.30 years, and for 2008–2020 this has increased to 2.97 years, so we see a slowdown in the by-core transistor counts. This is even more so for the uncorrected transistor counts: here 1995–2007 has a doubling time of 1.36 years—still near the Moore estimate[10]—and 2008–2020 has a doubling time of 2.79 years, almost twice that of the earlier interval.

We can perhaps infer that maximum design complexity, i.e. the most complex single processor core that can be designed by human minds, will be soon reached.[11] But even the total transistor count shows a similar slow-down of factor two which may indicate an upcoming exponential growth limit.[12]

One alternative interpretation of Moore's law is in terms of transistor density (i.e. $\frac{transistor\_count}{area\_in\_mm^2}$). In Fig. 2 we see transistor densities from 1995 to 2020 for CPUs, GPUs, a few deep learning engines[13] and some estimated densities derived from production technologies at the time they were introduced.[14] For 2020, we are already at 5 nm technology, but we do not have data on empirical densities. Still, since a silicon atom has a diameter of about 0.5 nm we are only a factor of 10 away from the physical limit of this technology. And since in 2007—about 13 years ago—we had 45 nm technology which is roughly ten times larger than 5 nm, this means just another 13 years to go before we reach physical limits, albeit in the unlikely case the doubling time does not increase further.

It also can be seen that CPUs and GPUs follow roughly the same curve, and are thus limited by the same constraints. From this graph we can again estimate doubling times on CPU data[15]: from 1995 to 2007 the doubling time is 1.78 years, and from

from 2006) where cache transistors are estimated at 1.5 billion by our method, but are actually 1.656 billion out of 1.72 billion (96.28%). We had to use this ad-hoc method as for most processors the number of cache transistors was never reported while cache memory sizes were readily available.

[10] Which is sometimes quoted—and used for long-term planning in the chip industry—as 1.5 years or 18 months.

[11] It may be argued that it has already been reached according to recently found bugs in almost all modern processors, see [22].

[12] GPUs contain millions of similar relatively simple cores and their transistor count per core would thus be much smaller in this graph, outside the shown range (data not shown).

[13] Only three datapoints.

[14] These are not intended as maximum *achievable* densities, as the technology will—if possible—be improved over time and may give slightly higher densities later, and these densities are also not known for many actual production technologies—only Intel reports some results. So please do not be confused by the seemingly impossible results in 2004, 2006, 2008 and 2010.

[15] We did not use GPU data since the number of datapoints was smaller while its shape is very similar to the CPU curve.

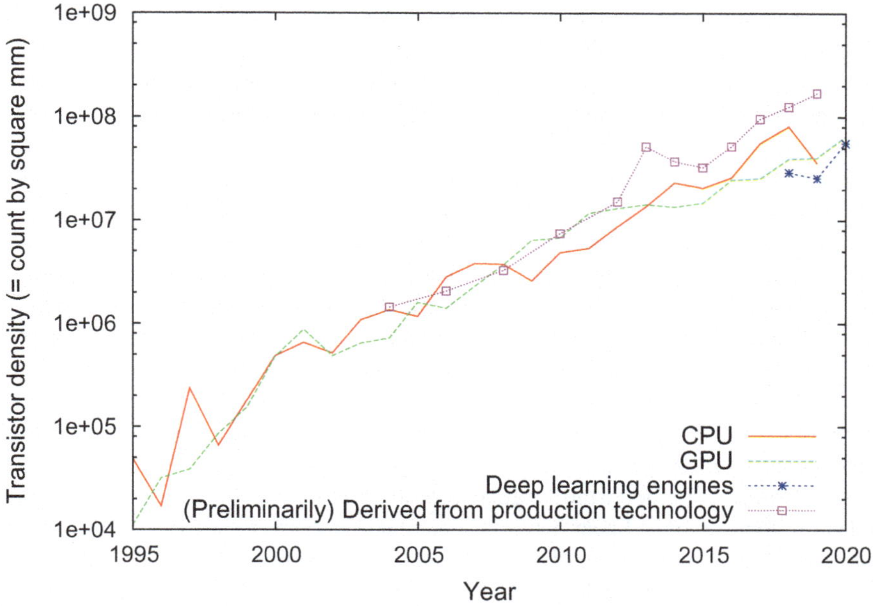

Fig. 2 Transistor density ($\frac{transistor\_count}{area\_in\_mm^2}$) between 1995 and 2020

2008 to 2019[16] the doubling time was 2.36 years, corresponding to a one magnitude increase in about 8 years. Again the doubling time has increased for the latter interval. So although the absolute theoretical limits are still 8–13 years away, we already see a reduction in the exponential magnitude, where according to singularity proponents we should be seeing a speed-up. Other estimates on when Moore's law will likely stop working are even earlier at 2025 [33, 34], which is consistent with an additional slowdown in the near future.

It has sometimes been mentioned that Moore's law has become a self-fulfilling prophecy, and it is true that semiconductor firms have used it for planning (as doubling in 18 months, see [30]). In fact the reliance on transistor counts for long-term planning is already reduced, amply demonstrated by the fact that Intel stopped reporting transistor counts for their processors in 2017. That there is a challenge to actually fill the technologically available transistor space can also be seen by the fact that from the 2016–2020 top processors, all but one (80%) are a systems-on-a-chip (SoC), and thus include other simple components otherwise part of the mainboard or chipset; and three (60%) include graphical processing units (GPUs) on-chip. This may indicate that we already have too much transistor space to usefully—and profitably—use for mere CPUs.

Additionally, as we noted earlier these observed exponential effects have been obtained by diverse overlapping techniques: from 1975 onward by reporting maximum-

[16] No data point existed for 2020 at the time of writing this paper.

complexity rather than minimum cost chips, from 2000 to 2010 by adding relatively simple cache memory, in 2005–2020 by successively increasing the number of identical cores on each die. There is no single underlying process to account for even the by-core transistor counts, so biological metaphors of an uniform growth process fall quite short.

Also, the success of the semiconductor industry has been heavily dependent on external factors, which can be seen by studying the history of the semiconductor industry. This industry faced several challenges, where they were in each case saved by external factors providing a large market for their chips—first for calculators and digital clocks, then for mini- and mainframe-computers, in the mid 1980s for the IBM PC and Microsoft with Personal Computers, and in the 1990s for the World Wide Web, which exploded the hard disk and memory market and also created the need for new powerful processor architectures to handle images, sound and highly compressed video [30]. If any of these events had not happened, Moore's law would have already been broken.

Even worse, as we noted the Singularity requires that costs to achieve this exponential speedup are at most linear in transistor count or density. However, the investment for a new semiconductor factory has also been increasing exponentially—doubling every four years—and reached US$ 14.3 billion in 2015. This is known as Rock's law, or Moore's second law [35]. In fact, already in 1979, Moore noted that the man-hours per month required for integrated circuit production were also growing exponentially. That all the effort put into making chips faster using cheap cache memory, pipelining, out-of-order execution, branch prediction, and other methods to increase performance has not been without issues can be seen in the recently found bugs such as Meltdown and Spectre [22]—an indication that we can build things we do not truly understand until decades later and that current developments costs, already quite high, are still underestimated.

3.1 Quantum Computing

Quantum Computing (QC [36]) is sometimes mentioned as an alternative to classical computing which may allow higher packing densities. At first glance it looks promising, since it uses *qubits* which can store quantum superpositions instead of 0-or-1 bits. This means a quantum computer with n qubits can store 2^n values and can also run reversible operations on them in parallel. However, the obtained results stored within the qubits cannot be read out directly since each measurement collapses the wave function and forces the qubit's value to either 0 or 1 which depends stochastically on the collapsed superposition. So for an useful algorithm as few measurements as possible must be performed to reconstruct the relevant parts of the collapsed wave function, restarting the quantum computation after each measurement. If this can be done successfully, quantum computers *may* be able to compute some functions exponentially faster than classical computers. Such *quantum supremacy* has however not been conclusively demonstrated and would in any case only be limited to a small set of algorithms proposed for quantum computers. For initializing the qubits, executing operations, and reading out the results classical computers are needed.

Quantum computers can potentially simulate normal computers. However, in this case each qubit corresponds to a classical bit and there is no advantage over classical computing w.r.t. transistor packing density.

It should also be noted that despite the potentially much higher packing density for temporary storage (qubits), the quantum gates themselves can—in almost all currently proposed quantum architectures—not be made smaller than a single atom, limiting the transistor packing density in a similar way as for classical computing.

3.2 Spintronics and Photonics

A combination of spintronics and photonics could achieve much faster computers running at terahertz frequencies—about 250 times faster than current CPUs—by combining information storage via magnetic spin and purely photonic gates. This is an active research area [37], but no working system with logic gates *and* memory has been proposed as of now. Even if such a system already existed, major additional minaturization would be needed to build a chip competitive to classical designs. If we also managed to shrink logic gates and memory units down to single atoms, it would put us at a point about 30 years beyond Moore's law. However, this would need major breakthroughs in this field over the next ten years and is therefore rather unlikely. Also, IBM [38] has already demonstrated a 100 GHz transistor in 2010, so higher clock rates may also be achievable using more classical transistor designs, negating this speed advantage.

3.3 Reversible Computation

Classical computation works irreversibly and has therefore a lower bound on heat dissipation due to entropy increase [39] which will become more relevant the closer we get to single-atom transistors. One possible solution for this is to use reversible computation which does not generate heat. This can be done by using reversible gates. However, reversible computation is inefficient in the sense that many bits must be preserved to enable reversibility which do not contribute to the actual computation and thus reduce the effectively usable transistor packing density. Also, if reversibility cannot be completely obtained, it may be necessary to run a reversible computer very slowly to limit heat dissipation. Such reversible computers are needed to drive quantum computing as even the small heat dissipation from the erasure of a single bit may destroy the fragile quantum states on which the operation of the quantum computer relies on. It is clear that such designs will always be slower than corresponding designs using irreversible computation and appropriate cooling.

4 A Case for History

In light of these negative results, why is the myth of the technological singularity still so compelling? To better understand this, let's take a look at AI history, starting with the precursor of AI—Cybernetics. This field is similarily contradictory and emotionally burdened as the singularity discussion, and retained its own persistent myth, as this quote shows:

> One pattern is spiritual. Not always, but often, the machine has become a godhead, an idol... Science created a totem. The machine became the avatar... A second powerful pattern is contradiction... Automated factories would free workers from undignified drudgery, yet deprive them of their dignity... Computers were dumb and could be hacked by teenagers, yet they could outsmart humans... More networked computers would lead into a "dossier society" of ubiquituous surveillance - and enable anonymity and a freer and better political order. Networked information system would make nations more vulnerable and more fragile than ever, and networked command-and-control systems would make their armies more dominant and more lethal than ever. Machines would be future society's hard-charging overlords and its soft underbelly. The myth [of cybernetics] hides these contradictions and makes them acceptable. (Rid [40], p. 348f, quoted with friendly permission by the author)

The myth of the upcoming technological singularity is very similar in spirit to the myth of cybernetics. In a way, the discussion on the singularity reflects earlier cybernetics discussions now only known to students of AI history.

In the research proposal for the Dartmouth meeting in 1956—where the term *Artificial Intelligence* was coined—it was proposed that "a 2 month, 10 man study of artificial intelligence be carried out" and it was concluded "We think that a significant advance can be made in one or more of these problems if a carefully selected group of scientists work on it together for a summer." [41]. Now, in 2021—65 years later—it is clear that the scope and breadth of the task has been severely underestimated by many orders of magnitude.

But Artificial Intelligence actually starts a bit earlier than that. We can see it already starting in 1943 with McCulloch and Pitts [42], who proposed a boolean circuit model of the brain.[17]

In 1950, Turing [43] reformulated the question of whether machines can think to whether machines can behave intelligently, and proposed an operational, non-constructive and non-reproducible test for intelligence—the Turing test. He suggested major components of AI: knowledge, reasoning, language understanding and learning, and anticipated all major arguments against AI. Nowadays, Chatbots based on text-mining terabytes of chat room logs are often judged intelligent by non-experts [44], however they still have practically no chance to fool an experienced AI researcher.

Early AI successes like Samuel's checkers program [45] which could beat human players and was trained only against itself, Newell and Simon's Logic Theorist [46] who found more elegant proofs for some known mathematical theorems, and Winograd's Blocks world [47], which could answer questions and act in a simplified blocks

[17] See Sect. 8 for an overview of different brain models that have been proposed in known history.

world with the abilities of a three-year old child have initially confirmed the optimism at the Dartmouth meeting.

However when we look closely with the benefit of hindsight, we note that—for example—checkers has a very simple structure that makes the simplistic approach work where it failed with more complex games such as Chess which was only solved many decades later. Another example is Backgammon, where a temporal-difference learning algorithm [48] achieved grandmaster-level performance with just a handful of heuristics. It was later found that the game's structure fit well with the coevolutionary learning paradigm [49], which explained most of its success and why temporal difference learning did not work on many other games. The blocks world AI actually only understood very specific simple sentence patterns. Even slightly different formulations or typing errors make the system fail to understand—it is extremely brittle like most early AI systems. Now, 65 years later, we are finally able to build systems which understand written and even spoken speech far more robustly, but it has been extremely hard to get here.[18]

One of the reasons why it has been hard was computational complexity. Most hard problems get exponentially harder with larger problem sizes, meaning that small instances (toy problems) can be easily solved, but even slightly larger problems take years, and realistic problems may take longer to solve than the remaining lifetime of the universe.[19]

The solution proposed around 1969–1979 was knowledge-based systems, resulting in the birth of the expert systems industry. However, knowledge as stored and processed by machines is very different from knowledge stored and processed by humans. For example, contradictory statements make a knowledge base completely useless,[20] which makes it exponentially harder to build larger knowledge bases, as they have to be completely free of even the smallest contradiction. So this problem—exacerbated by overarching promises—later lead to a bust in the expert systems industry: the so-called *AI Winter*.

From 1988, there was a resurgence of probability, increase in technical depth, and the birth of Nouvelle AI: Artificial Life, Genetic Algorithms, and soft computing such as Fuzzy Logic. From 1995 the beginning of the Agents metaphor, a stronger focus on embodiment—another thing which is coupled strongly with human intelligence—and an increasing number of real-world applications was observed.

The Dot-Com crisis from 2000 reduced investment into computer technology for some time, so in the first decade of the 21st century, not much progress was made. Still, in 2005 Stanford's Stanley [51] won the DARPA Grand Challenge, driving autonomously 131 miles through the desert without using a camera. Laser-range LIDAR sensors were used throughout and almost all parts of the system were trained

[18] Compare e.g. the simplicity of TD-Gammon learning with AlphaGo [50] w.r.t algorithmic complexity and the amount of computational power needed to train the system.

[19] According to the big crunch theory. If our universe were to expand indefinitely, we would still at some point run out of energy, but it would be much later.

[20] Known as *ex falso quodlibet*: from false [knowledge] anything can be derived.

and tuned using machine learning algorithms, showing a clear advantage over the handcrafted systems used by competitors.

In 2010, Microsoft introduced the Kinect 360° motion sensor, which had been trained using Random Forest classifiers [52] to recognize human figures from depth camera[21] recordings. In 2011, IBM's Watson [53] beat the two greatest Jeopardy! champions.

Deep Learning took off in 2012 with the seminal paper by Hinton et al. [54]. This was due to several factors—wide availability of software frameworks, cheap GPU hardware due to the successful gaming industry, and better learning algorithms—which we will go into more detail later in Sect. 5. In 2015 Tesla announced a software update to enable self-driving for their electric cars on freeways. Their system was end-to-end trained using Deep Learning, albeit using only cheap 2D cameras and radar, and is therefore not as robust as LIDAR-based self-driving systems. In 2016, Google DeepMind AlphaGo [50] beat 9dan Go champion Lee Sedol 4:1. In 2017 AlphaZero [55] demonstrated learning to play Chess and Shogi (chinese chess) at champion level in days. Rational solvers have been optimized to such an extent that large proofs can be generated in days, such as the proof that the 5th Schur number S(5) equals 160 [56]. This proof is 2 petabytes large, and needed 14 CPU years computation time. In 2019, AlphaStar [57] won in the real-time strategy game Starcraft II 10:1 again human expert players. In 2020, AlphaFold [58] won the biannual Critical Assessment of Protein Structure Prediction competition by predicting the 3D structure of proteins much better than the closest competitor. So it is clear that Deep Learning dominates AI research at present. But how did this happen?

5 The Rise of Deep Learning

No criticism of the singularity would be complete without a section on Deep Learning. The history of deep learning started with the already mentioned work of McCulloch and Pitts [42] who presented a simplified model of nerve cells, which computed a thresholded step function on a weighted sum of inputs. Rosenblatt [59] introduced a formal definition for such a system, coined the term *perceptron* and demonstrated a learning algorithm that could learn simple input-output functions such as AND, OR and NOT in a single perceptron.

Minsky and Papert [60] incidentially proved in 1969 that a single perceptron could not learn the XOR function—or any other not linear separable function. Although Rosenblatt could demonstrate—as McCulloch and Pitts [42] had done before—that a combination of several perceptrons could learn XOR, the lack of a suitable training algorithm for such combinations of perceptrons still lead to an almost complete stop in research on neural networks for about a decade. The perceptron remains in its original

[21] A depth camera returns pixel arrays similar to a normal camera. However the pixel values correspond to the measured distance of each pixel from the camera plane rather than brightness or color. These distance measurements are usually obtained via active sensing.

and modified forms the core unit for neural networks, including convolutional and deep learning networks.

Only when Werbos [61] independently reinvented concepts already known by Kelley [62],[22] and proposed the backpropagation learning algorithm for training networks of interconnected perceptrons arranged in layers (multi-layer perceptrons, MLP[23]), research continued albeit at a relatively low level. Rumelhart [64] soon demonstrated that this learning algorithm led to useful internal representations for the internal layers.

LeCun [65] introduced a new form of neural network which was inspired by the visual system[24] and arranged the perceptrons in their natural 1D, 2D or 3D order and introduced two new types of layers[25]:

- Pooled or subsampling layers, which compute global statistics, such as maximum or average over a window of perceptrons from the previous layer, just like a simple filter.
- Convolutional layers, where weights are shared between perceptron units, i.e. each point of the next layer computes a sum with the same weights, but on a different window of the previous layer—a filter or convolution in image space.

Such a network was trained to recognize handwritten digits for the US postal service and similar networks are still used by many OCR systems. Lastly the seminal paper by Hinton et al. [54] to which all the fields's major research labs and researchers contributed lead to renewed interest in this field and—due to the new focus on networks with many levels, although it also introduced new layer and interconnection types and perceptron activation functions—it was renamed to Deep Learning.

So why did it take Deep Learning so long to take off? Earlier, it was generally believed that neural networks were very unstable learning systems and not suited for a wide range of problems. It was believed that overfitting and getting stuck in local minima would be a major problem for many tasks. Also, it was considered infeasible to obtain feature sets competitive to handcrafted features, especially for image and audio processing which had been researched very intensively up to this point. The lack of publicly available source code to train such systems was also a problem. For example, we replicated work by LeCun [65] in [66], but the source code used to train the convolutional network model was only available under a cumbersome license, and heavily tailored for the task of handwritten digit recognition, and thus could not be used for other projects.

---

[22] Tragically, this work would already have been available when Minskey & Paperts book was published. However since it was a completely different application and research field, this was not noticed until much later.

[23] A second way to solve the XOR-problem for perceptrons is to regularize the model space by means of maximum margin hyperplanes (also making the solution unique) and introducing the kernel trick—expanding the input feature vector with base functions. This lead the the field of Support Vector Machines (SVM, [63]).

[24] It however ignores the 90% LGN feedback connections which we will mention in Sect. 8.

[25] Previously, only fully connected layers were used where each perceptron within one layer is connected to all perceptrons within the next layer.

To enumerate more systematically, the main reasons why Deep Learning took off were the following:

1. Moore's law and the corresponding increase in single and multi-processor performance since 1970. This was especially important to Deep Learning, as classical (human-written) machine learning algorithms are very seldom highly scalable[26]—i.e. they profit linearily from *faster* processors, but much less from *more* processors. Now that Moore's law is mostly driven by integrating even more parallel processors, classical code profits much less from the speed-up than Deep Learning. In fact, Deep Learning training algorithms are almost infinitely scalable and can be run in parallel with one processor per perceptron unit, of which there can be several million albeit communication costs will at a certain point begin to dominate, limiting scalability.
2. The resurgence of the vector processor in form of graphical processing units (GPUs).[27] This factor was driven by the gaming industry in its goal to create ever better computer game graphics on similarily improving monitors (in terms of resolution and update frequency). Training and evaluation of Deep Learning networks basically involves computing relatively simple, very similar operations on a large set of units. A normal CPU can of course perform these operations, but is optimized for a more diverse workload and normally slower than a specialized GPU for such simple essentially vector operations. It turned out that GPUs can easily be used for Deep Learning and initially speeded up training by one—and later two to three—orders of magnitude, and in fact there are now specialized processors optimized for Deep Learning tasks that obtain even higher speedups.
3. Once a large set of research groups applied these methods, it became obvious that the perceived disadvantages (instability, local minima) were not significant and could be overcome relatively easily. In fact Deep Learning systems soon dominated visual recognition, image classification and speech recognition tasks within a few years, thus confirming that they are able to create internal feature spaces (represented within the first few layers after the input layer) that are superior to handcrafted human feature spaces in many tasks. This does not mean that Deep Learning system are without issues—failure modes as described in [68][28] and [69][29] indicate that their models are quite different and less robust than the ones human brains use.
4. New and improved training algorithms such as Adam [70] reduced both training time and the risk of getting stuck in local minima, which had previously been a significant problem in training deep learning networks. Initializing the model

---

[26] Except perhaps for Random Forest [67], which we've applied in unpublished research and found to scale very well.

[27] If you remember the large Cray computers—yes, these were vector processors as well and they could have been used for Deep Learning already in the 80ies, forty years ago as of now.

[28] Arbitrarily changing object classifications and confidences by specific changes in background colors for input images.

[29] Arbitrarily changing object classifications and confidences by adding specific noise to the input image.

weights not randomly but with statistical methods (e.g. PCA) also adressed these issues and furthermore speeded up training time significantly. Both made it possible to train larger and more complex network models.

5. With Torch (2002), Theano (2007) and especially Tensorflow (2015) [71], a large set of deep learning frameworks with GPU support became mature and widely used. Although Theano has ceased development in 2017, and Torch in 2018, pyTorch—a fork of Torch written in Python—is still in active development as of December 2020.
   Especially Tensorflow, which contains a large set of samples and pretrained models including training code and in most cases the data needed to retrain the model itself—combined with the interface in the easy-to-learn programming language Python—has made Tensorflow very popular. From version 1.5, models could be converted to Tensorflow Lite to run on mobile phones and embedded platforms such as the Raspberry Pi. All these frameworks—most of which are available under a permissive license such as MIT or Apache 2.0—made applying the corresponding learning algorithms much simpler, and led to a corresponding increase in researchers using deep learning, and therefore also in the number of applications.

But does all this mean we have already—or will shortly have—achieved human-level intelligence, i.e. an Artificial General Intelligence (AGI)? Sadly, no.

6 The Fall of Deep Learning

One major reason why Deep Learning is not an Artificial General Intelligence (i.e. a human-level AI) is that our training algorithms create models that have completely different—and far more brittle—failure modes than human intelligence, amply demonstrated by systems such as DeepFool and ColorFool [68, 69]. So the learned models are quite different. This alone however could simply be an artefact of the network architecture. However, it is also quite inconceivable that even relatively simple learning algorithms such as Backpropagation—to say nothing of Adam—could be actually used within our brains. The learning algorithms are much too different. Also, the nerve system model of Deep Learning still basically reflects the perceptron and is by no means a biologically plausible model as this incomplete list of open issues shows.

1. It is unclear how the goal values could be transported through all neural layers to compute the error feedback. Also, despite decades of experiments, no backpropagation error signal was ever observed.
2. Biological neurons communicate via binary spikes and not via numerical values
3. Synchronization between neurons—an extremely important mechanism used in the brain—is not modelled in almost all Deep Learning networks.

4. Backpropagation would need perfectly synchronized discrete steps which is not possible as biological nervous systems do not have a sufficient precise global clock.
5. The weights for the backpropagation connections would have to mirror the weights of the corresponding forward connections. There is no known biological mechanism which could ensure this.
6. Backpropagation would need to know the non-linear derivates of all neurons in the higher layers. Again, there is no known biological mechanisms which could—even potentially—compute these values.

It is clear that Deep Learning reflects one deeper principle on how neural networks build complex models, but it is also similarily clear that it is just a small part of the whole picture.

Additionally, the original inspiration for convolutional neural networks came from the human visual system—however the 90% feedback connection to the LGN nucleus from almost all other brain areas were completely ignored. But such massive feedback loops may be essential for robustness (more details see Sect. 8).

The DeepMind series of AlphaGo [50] and AlphaZero [55] models is perhaps best suited to show the disadvantages of Deep Learning from a different view point: First, these system can only be trained by self-playing resp. by creating extremely large data sets (sometimes implicitly) as millions to billions of samples are needed. No human player can generate even a small part of the training data these systems need to get up to speed. This also means that if one player notes a way to beat the system—easier in perfect information games such as Go, Chess or Shogi, since there are no random fluctuations in the environment—it is always possible to beat the system in exactly the same way, and there is no easy fix for this.[30] It's possible to force the system by retraining to choose another path at a critical juncture, however this might weaken overall performance, and in any case there will always be other weaknesses to exploit. What cannot be done is to train the system to play well against a specific player since—as noted above—even if this player played Go games their whole life, and all games were stored and trained, this would not even make the smallest difference.[31] Especially for AlphaStar [57] (online first-person gaming) it was mentioned that initially the model used recorded human player movements as otherwise the exploration state space would simply have been too large to find meaningful movement patterns. However note that children and teenagers have no problems learning to play this game without such information in single player mode.

Realistic estimates towards when Artificial General intelligence will be available tend towards several centuries. As mentioned Hanson [17] (p. 63) estimated 4–8 centuries should Moore's law reduce just by a factor of two but still keep exponential over the whole time period, which is actually very unlikely as we will hit strong physical limits in the next decade. Since the beginning of AI research in 1956, the

[30] We may assume the creators of AlphaGo are aware of this issue and perhaps it informed their decision not to make AlphaGo publicly available.

[31] This was clearly mentioned at one of the press conferences after the AlphaGo win, and should of course be obvious considering the large amount of data such systems need for training.

median forecast of the duration of time until human-level abilities would be achieved in AI research was always around 30 years. Needless to say, all these forecasts were wrong.

So, sadly, although Deep Learning is a big step towards understanding intelligence —or more precisely, how extremely simplified pseudo-nervous system can be trained using biologically implausible learning algorithms to solve tasks that for humans require intelligence—but not an AGI at all. Many further significant advances would clearly be necessary to achieve this.

7 Singularity Contradictions

Should we somehow still manage—by serendipity, sheer luck or perhaps as a download from an extinct extraterrestrial civilization – to create an Artificial General Intelligence, it is by no means assured that it will achieve superintelligence. There are quite a few somewhat obvious contradictions within the story of the technological singularity which would also apply to any AGI (including ourselves, of course).

7.1 The Myth of Better Introspection

It is seldom questioned that an AGI would be able to bootstrap itself to higher intelligence quite easily. However, why would an AI at the same intelligence level as us have better insight into its internal structure? Why should its mind work parallel and not serially as ours, limiting the obtainable speedup as well as self-reflection? We note that although the human brain is a massively parallel system, the human mind is still only able to do one thing at a time and is very bad at multitasking. It is not obvious why such a constraint should not be a general constraint of all minds, perhaps driven by the necessity for limited self-awareness and an unified stream of consciousness.

A superintelligence is expected to be able to fully understand humans. But how to get from a human-level Artificial General Intelligence to a superintelligence when it cannot improve itself much better than a human? The exponentially increasing speed of self-improvement, so essential for the technological singularity, is not likely to hold at any intelligence level. Otherwise programming code would not still be written by humans. It is for example possible to train Deep Learning networks with a few millions lines of C code. They will then output syntactically correct code with few errors (mostly concerning assignments to previously undeclared variables) that is completely meaningless. Without a guiding mind and intentionality behind it, even extremely subtle abilities are essentially useless. And a true AGI written e.g. in C that would be as proficient as the best human C programmer would still be no better suited to rewrite itself to become smarter—similar as we humans do not have a privileged insight into the workings of the 10 billion neurons of our brain.

Even if a superhuman intelligence could be created somehow, it would potentially be able to improve lesser human-level AI systems—perhaps even to its own level—but it is rather unlikely that it would be able to indefinitely improve itself at ever increasing rate. As any programmer knows, radical speedups can sometimes be obtained, but the law of diminishing returns sets in and at some point the effort to make a program go faster is simply not worth the obtainable speed-up. In many cases, optimizations are even detrimental and reduce performance. It is unclear why this should be different for the extremely complex operations normally associated with high-level cognition. If anything, one would expect even steeper diminishing returns here. And improvement in hardware is time-consuming and costly. A new semiconductor plant takes billions of dollars to set up and only runs cost-effectively after several years of extensive experimentation and optimization. To believe a superhuman intelligence can cut through all these issues, and ignore physical reality and all manufacturing constraints in building its successor is only a pipe dream. Especially since it is well-known that building robots that are as flexible and versatile as the human body is extremely hard, as anyone who has ever built robots can attest to. Despite decades (soon to be a century) of research the state-of-the-art bipedal robot apocalypse could at present be fended off at the top of stairs with slingshots. In fact embodied robots instantly seem much more intelligent than their simulated counterparts even if they are not, simply by utilizing the randomness of real environments, and it is unclear whether an intelligence could ever be created without some kind of grounding in physical reality—which would again preclude superfast self-improvement.

Concluding, we find it increasingly implausible that even a superhuman mind could bootstrap itself to a significantly higher intelligence level, and more so that it could do so at an ever increasing rate (as in Kurzweil's [12] law of accelerating returns). So even if a superhuman AI would magically drop into our hands, it would not cause a singularity. And it would have to drop, since despite 65 years of research we have actually no idea how to build one, or even just an AGI.

7.2 The Myth of Continued Exponential Economy Growth

Assuming an economic doubling time of 15 years (assumed by [17]), a century means about seven doublings, meaning that even in one century the economy would be $2^7 = 128$ times larger than it is now. As we mentioned, AGIs—surely a necessary precursor of the singularity—are at least two and more likely 4–8 centuries away. Historically, there has been a strong correlation between economical growth and energy consumption, but even a small poportion of a 128-fold increase in energy consumption would probably change climate to such an extent that human survival is not longer feasible.

The singularity story mainly counters this with *superexponential dematerialisation*, i.e. the theory that it is possible to obtain similar growth rates while using ever less energy and matter. This implies not only an uncoupling of energy consumption and growth, but an inverse correlation—something that was never seen in the past of the industrial revolution for the last 200 years.[32] But as the necessary technologies are at least several decades in the future—if they can be made to work at all—this will come too late as only a single doubling in energy consumption (and thus CO^2 emissions) will be enough to change the biosphere in a way that makes living on this planet no longer feasible for the number of people who are currently living here. Much more likely is a radical reduction of growth rates in energy consumption: for example Sharma [72] quotes a global growth rate of energy of 0.8% for 2016–2030 and 0.1% for 2030–2060. This would mean diverting time into more efficient processors which need less energy, and reducing the expected growth rate in computer performance further.

For comparison, the human brain consumes about 50W of energy and generates this energy in a completely sustainable way, while Google consumed 10.6 Terawatt hours in 2018, which is up from 2.86 Terawatt hours in 2011, corresponding to a growth of 20.58% per years, or a doubling time of 1.66 years (20 months), even faster than the doubling time in transistor count. It is clear this cannot be sustained for even one additional century.[33]

In fact Gordon [73] points out that measured economic growth has slowed around 1970 and slowed even further since the financial crisis of 2007/2008, and also argues that the economic data show no trace of a coming Singularity as imagined e.g. by Good [5] already in 1966. So the growth observed due to Moore's law seems to be the exception rather than the rule, but even this is slowing down as we already showed.

7.3 The Myth of Recursive Self-improvement in Research

Another reason why the arbitrary speedup so desparately needed for the technological singularity is a mirage: The reason why research follows phases is that first a critical set of observations must be present before a theory can be formulated. This process cannot be arbitrarily sped up—e.g. new physical measurement devices must be designed, tested and applied. For example, in physics before Quantum Mechanics, the consensus in the field was that basically most of physics was solved, and only a few strange quirky observations remain which would soon be resolved—such as black body radiation, which is predicted by classical theory to be infinitely high when integrated over all wavelengths. This was called the Rayleigh-Jenes law or ultraviolet catastrophe. Quantum mechanisms quantized the photon output and thus obtained results consistent with observations, but most of physics had to be reevaluated in light of these new assumptions.

---

[32] It is also quite unlikely that it ever happened before in human history, seeing that for most of history almost all people lived near subsistence levels.

[33] Google claims to use 0.3 Wh for the average query. With this amount of energy an average human can think for 21.6s, which *might* just be enough to remember what you were searching for yourself instead of googling it.

In fact Alston [15] found that research funding gives far less than linear returns. It is inconceivable that any kind of intelligence is somehow completely immune from these fallacies and pitfalls, especially since the requirement for recursive ever-faster self-improvement prevents its grounding in physical reality, which may be an essential condition for intentionality, consciousness and human-level intelligence.

8 Mind Is Always the Highest Available Technology ... and Always Will Be

It is part of a long-standing tradition in human history to take the newest technology and propose that humans—or the human mind—work that way. For example, Galen of Pergamon (129–216 CE) thought that the psychic pneuma (cerebrospinal fluid) who circulate in the ventricles (liquid-filled holes in and around the brain) is conducted along the nerves (thought to be hollow) and effects all psychic phenoma such as sensation, elaborating thoughts and movements—i.e. the mind [74].[34] This was at a time when the first distillation experiments were conducted, and parlour tricks such as boiling wine and igniting the alcohol vapours inspired the idea that invisible influences may be present in liquids.

In 19th century, Victorian England, Huxley [75] proposed that human react as automata in all our apparently free decisions. This was probably inspired by ingenious automata such as [76]—an automaton that could write three different poems and four different drawings; [77]—a mechanical duck that could quack, muddle the water with its bill, drink and eat and seemingly defecate (although the last part was faked); and [78]—an automaton that could reproduce natural speech, as well as countless others including numerous fakes such as the Mechanical Turk. The astonishing performance of these devices led to Radical Behaviourism where the metaphor of humans as automata was upheld for decades despite ample evidence to the contrary.

We already mentioned McCulloch and Pitts [42] who proposed a boolean circuit model of the human brain at the beginning of the computer era, where boolean circuits were the highest available technology.

Quantum mechanics is the theory that brought us—among other things—extremely fast computers, lasers, DVD and bluray players, LEDs and the atomic bomb—and can thus also be considered the highest technology a few decades later. Therefore it should not surprise us that Penrose [79] proposed a quantum-mechanical mechanism to explain the mind, which has been taken up by many other researchers (for an overview see Adams [80])—although admittedly many of their hypotheses cannot be falsified.

In the 21st century, we now have Deep Learning and again it is assumed that the brain works in a similar way. However, this is a fallacy for several reasons. For one, the best networks are trained used ground truth data in a teacher/pupil setting,

[34] Others such as Aristotle saw the brain as mechanism intended for cooling blood, and instead the heart as the seat of soul and mind.

while most of the complex work of the brain (such as learning head/eye coordination, segmentation of world images, walking, learning to speak, exploration of unknown environments, ...) is done with no or almost no explicit external feedback. We simply do not have unsupervised learning algorithms that work as well as the supervised learning algorithms commonly used. Secondly, trained networks can be easily provoked to make mistakes, either by adding specific noise into the input images [69] or by specific changes in background colors [68]. Human image segmentation is not susceptible to either attack, in practice the changes—especially of the latter—are so subtle that they are hardly visible even when the original and modified images are shown side-by-side, but they still completely confuse DL systems. Optical illusions in humans are on the other hand far more subtle and complex despite our eyes recording images of much worse quality.

A close look at the brain wiring might give us an answer as to why Deep Learning networks modelled on the visual system have these massive vulnerabilities: the lateral geniculate nucleus (LGN), where 90% of the retinal axons (who carry information from the eye) terminate, gets only 10–20% of its input from these connections ([81], p. 532). Surprisingly, 80–90% of its input connections originate in many other regions of the brain.[35] It is quite likely that these other connections modulate and adapt image processing yielding the practically perfect segmentation we effortlessly create based on rather bad "eye camera" input. However, if that is the case, to get similar performance, we must also emulate the rest of the brain—the task of image segmentation is then AI-complete, and thus needs an Artificial General Intelligence (AGI). Again, reasonable estimates place AGI at least two, and more likely 4–8 centuries into the future ([17] pp. 61–63).

9 Conclusion, Discussion and Outlook

We are not overly concerned that the exponential growth in computing performance will quite likely stop in the near future, as it will also take care of planned and unplanned obsolescence, the huge amount of electronics garbage generated by neglecting the actual possible lifetime of electronics devices, the abysmal conditions under which people prospect essential heavy metals such as lithium—the main component in rechargable batteries which are major parts of smartphones, notebooks, and electric cars—and the large amount of time we spend to reacquaint ourselves with new versions of software which are often less intuitive and slower than the old versions.

On the downside we will have to think more deeply about optimizing algorithms and creating—or training—newer, more efficient ones. But since we already have more computing power than we can sensibly put to use, that is not a large issue.

[35] Confirmed are connections to the visual cortex, superior colliculus, pretectum, thalamic reticular nuclei, local LGN interneurons, mesencephalic reticular formation, dorsal raphe nucleus, periaqueductal grey matter, locus coeruleus and the optic tectum (superior colliculus).

We can no longer expect our old systems and algorithms to get exponentially faster automatically. But we won't have to buy or rent new hardware every few years to realize these gains, which more than offsets these disadvantages.

So where do we go from here? We have taken the liberty of writing up the areas where—in our humble opinion—too little is done presently.

1. Currently, Deep Learning algorithms focus mainly on the supervised learning case. However, unsupervised and reinforcement learning in realistic settings—i.e. much less training data, transductive and transfer learning, and especially embodied learning on robots interacting with the physical world—is severely underrepresented. It is our conviction that simulations and offline-learning, albeit useful, can never replace the interaction with physical reality, and that this is the one area where Artificial General Intelligence (AGI) will—eventually—emerge (in 4–8 centuries). We would go so far as to propose that each AI researcher should be forced to build robots as part of his/her education.[36]
2. We have many frameworks for Deep Learning on sufficiently powerful computers. However, there have been only a handful of systems that allow learning directly on the robot where power consumption, runtime and computational power are severely restricted. We do not yet have the learning algorithms that are sufficiently fast to learn big networks in real-time on these platforms, but perhaps we can work on this. Theoretically, streaming data to a larger platform via Wifi can resolve this to some extent, however our experiences were mixed since latency is a large issue and cannot be completely controlled, also the transfer of uncompressed video frames[37] is more severely limited by latency than bandwidth [83], so the upcoming 5G standard which focusses mostly on bandwidth will not be much help.
3. Deep Learning algorithms are not at all implemented in a biologically plausible way. Although good results are often obtained, it is unclear how our learning algorithms could ever be implemented in an actual biological neural system. So it is quite clear than biological systems must use different—and clearly superior—learning algorithms. There is much work on how biological in vitro neural assemblies can be trained for tasks, but again the underlying algorithms are not known. For example, synchronization between neural assemblies—an important mechanisms for e.g. binding and attention control—is almost completely ignored in the Deep Learning community. It would be interesting to study the actual algorithms used by biological neural systems which after all have had much time to perfect their methods.

[36] Of course this will not always work—some roboticists sometimes get caught in these myths as well, e.g. [10]. So perhaps a psychological course on the dangers of antropomorphisation should also be mandatory.

[37] We note that the existing compression algorithms have been optimized to remove parts of images and videos not observable by the human eye and are likely suboptimal for a non-human visual system. At the very least, training should be done on exactly the same data which was originally collected as models trained on compressed and uncompressed data differ and can not be used interchangably [82].

A small model organisms, C. elegans, commonly used for ageing research, has a very small nervous system of only 302 nerve cells. Although its nerve cells are different from mammalian nerve cells,[38] if we could record sufficiently detailed traces of the complete nervous system, we may be able to decipher its workings. This organism has six known high-level learning behaviours, some of which can also be triggered in the adult specimen. It would therefore be a reasonable starting point for such an analysis, perhaps better suited than large biological neural assemblies as it is a fully embodied and biologically grounded organism. We have done some minor work on this organism in [84], but so far nobody has proposed a recording system that can reliably record all neurons of this organisms with sufficient temporal and spatial accuracy. Once this is solved, progress may be rapid.

4. The human brain uses about 50W of energy. However the capabilities of the human brain are so much higher than the best currently available platforms which used comparably amounts of power. There are several recent successes in AI e.g. AlphaGo [50], but the computers and GPUs needed to achieve these successes use approximately 6–10 orders of magnitude more power than the human brain. If nothing else, it would be interesting to focus more resources on researching energy-efficient computing systems and platforms. To some extent this has already begun. This point somewhat overlaps with 2.
5. All minds need self-representation, and it is quite unclear how even modest self-improvement at the highest level can happen without such a self-loop representation. However, presently most Deep Learning networks are either not recurrent (i.e. the do not contain loops) or they only loop inputs or single intermediate layers, not exhibiting the rich kinds of loops between many areas apparent in the human brain or other nervous systems. The reason for this is mainly limitations of our training algorithms and lack of suitable training data (see also point 1 on the necessity for more research into unsupervised learning). Hofstadter [85] has also proposed such loops as central to mind and intelligence.

These are the things we can do as small steps towards an AGI. However, serendipity may achieve what we cannot directly aim for. While we do not believe AGI will emerge spontaneously on the internet (as imagined in [86])—the internet, although rich in content and structure, is neither as complex as the real world nor does it have a sufficient set of strict rules that govern possible interactions and serve as seeds for symbolic computation—it could perhaps emerge in a given sufficiently complex computational structure that also includes interaction with physical reality (as imagined by Hogan [87]). The same author has also written a very insightful critical book on Artificial Intelligence [88] which—although somewhat dated—still contains many valid criticisms.

But let's not cite only science fiction authors on this: Austrian Nobel Prize winner and physicist Erwin Schrödinger notes in the epilogue of *What is Life?* and also to some extent implies in *Mind and Matter* that all minds must be implemented at the lowest level of physical reality, i.e. "I ... am the person, if any, who controls the' motion

---

[38] They do not have Ca_{2+} channels and therefore do not exhibit action potentials.

of the atoms' according to the Laws of Nature." ([26], p. 87). A candidate mechanism, *Orchestrated Objective Reduction*, has been proposed by [79] and defended in [89], however it is quite implausible and has been severely critized. Still, if we follow Schrödinger on this, mind must be found at quantum levels or below. So perhaps we will someday obtain valuable information from the unlikely direction of fundamental physics research as well.

References

1. Vinge, V.: Marooned in Realtime. Macmillan (2004)
2. De Abaitua, M.: The Red Men. Hachette, UK (2013)
3. De Abaitua, M.: The Destructives, vol. 3. Watkins Media Limited (2016)
4. Ulam, S.: Tribute to John von Neumann. Bull. Am. Math. Soc. **64**(3), 1–49 (1958)
5. Good, I.J.: Speculations concerning the first ultraintelligent machine. In: Advances in Computers, vol. 6, pp. 31–88. Elsevier (1966)
6. Lem, S.: Golem XIV. Wydawnictwo Literackie, Krakow (1981)
7. Vinge, V.: First word. Omni **10** (1983)
8. Vinge, V.: Technological singularity. In: VISION-21 Symposium Sponsored by NASA Lewis Research Center and the Ohio Aerospace Institute, pp. 30–31 (1993)
9. Solomonoff, R.J.: The time scale of artificial intelligence: Reflections on social effects. Human Syst. Manag. **5**(2), 149–153 (1985)
10. Moravec, H.: Mind Children. Harvard University Press, Cambridge (1988)
11. Kurzweil, R., Richter, R., Kurzweil, R., Schneider, M.L.: The Age of Intelligent Machines, vol. 579. MIT press Cambridge (1990)
12. Kurzweil, R.: The singularity is near: when humans transcend biology. Penguin (2005)
13. Yudkowsky, E.: Three major singularity schools. Singularity Institute Blog. 30 Sept 2007
14. Sandberg, A.: An overview of models of technological singularity. In: The Transhumanist Reader: Classical and Contemporary Essays on the Science, Technology, and Philosophy of the Human Future, pp. 376–394 (2013)
15. Alston, J.M., Andersen, M.A., James, J.S., Pardey, P.G.: The economic returns to US public agricultural research. Am. J. Agricult. Econ. **93**(5), 1257–1277 (2011)
16. Hanson, R.: AI progress estimate. Overcoming Bias blog, 27 Aug 2012. https://www.overcomingbias.com/2012/08/ai-progress-estimate.html
17. Hanson, R.: The Age of Em: Work, Love, and Life when Robots Rule the Earth, 2nd edn. (paperback) edn. Oxford University Press (2018)
18. Mercier, H., Sperber, D.: Why do humans reason? Arguments for an argumentative theory. Behav. Brain Sci. **34**(2) (2011). https://doi.org/10.1017/S0140525X10000968
19. Nisbett, R.E., Wilson, T.D.: Telling more than we can know: verbal reports on mental processes. Psychol. Rev. **84**(3), 231 (1977)
20. Stumpf, E.: Förderung bei Hochbegabung. Kohlhammer Verlag (2011)
21. Naam, R.: Why AIs won't ascend in the blink of an eye–some math (2014). http://www.antipope.org/charlie/blog-static/2014/02/why-ais-wont-ascend-in-blink-of-an-eye.html
22. Lipp, M., Schwarz, M., Gruss, D., Prescher, T., Haas, W., Fogh, A., Horn, J., Mangard, S., Kocher, P., Genkin, D., et al.: Meltdown: Reading kernel memory from user space. In: 27th {USENIX} Security Symposium ({USENIX} Security 18), pp. 973–990 (2018)
23. Naam, R.: The singularity is further than it appears. Charlie's Diary (2014). http://www.antipope.org/charlie/blog-static/2014/02/the-singularity-is-further-tha.html
24. Ko, K.H.: Hominin interbreeding and the evolution of human variation. J. Biol. Res. Thessaloniki **23**(1), 17 (2016)
25. Darwin, C.: The origin of species. PF Collier & son New York (1909)

26. Schrodinger, R., Schrödinger, E., Dinger, E.S.: What is life?: With mind and matter and autobiographical sketches. Cambridge University Press (1992)
27. Sterling, B.: Schismatrix plus. Ace Books/Berkeley Publishing Group (1996)
28. Parker, M.P.: Intelligence and dysgenic fertility: re-specification and reanalysis. Chrestomathy Ann. Rev. Undergraduate Res. Coll. Charleston, **3**, 167–181 (2004)
29. Grace, K.: Algorithmic progress in six domains. Technical report. Machine Intelligence Research Institute (2013)
30. Tuomi, I.: The lives and death of Moore's Law. First Monday (2002)
31. Wikipedia contributors: transistor Count. https://en.wikipedia.org/wiki/Transistor_count (2020). Downloaded on 24 Nov 2020
32. Wikipedia contributors: Montecito (Processor). https://en.wikipedia.org/wiki/Montecito_(processor) (2020). Downloaded on 4 Dec 2020
33. Waldrop, B.M.: The chips are down for Moore's law. Nature **530**, 145 (2016)
34. Kumar, S.: Fundamental limits to Moore's law. arXiv preprint arXiv:1511.05956 (2015)
35. Meieran, E.S.: 21st century semiconductor manufacturing capabilities. Intel Technol. J. **1**(8) (1998)
36. Steane, A.: Quantum computing. Rep. Progr. Phys. **61**(2), 117 (1998)
37. Lalieu, M.L., Lavrijsen, R., Koopmans, B.: Integrating all-optical switching with spintronics. Nat. Commun. **10**(1), 1–6 (2019)
38. Lin, Y.M., Dimitrakopoulos, C., Jenkins, K.A., Farmer, D.B., Chiu, H.Y., Grill, A., Avouris, P.: 100-ghz transistors from wafer-scale epitaxial graphene. Science **327**(5966), 662 (2010)
39. Landauer, R.: Irreversibility and heat generation in the computing process. IBM J. Res. Dev. **5**(3), 183–191 (1961)
40. Rid, T.: Rise of the machines: the lost history of cybernetics. Scribe Publications (2016)
41. McCarthy, J., Minsky, M.L., Rochester, N., Shannon, C.E.: A proposal for the dartmouth summer research project on artificial intelligence, august 31, 1955 (Reprint). AI Mag. **27**(4), 12–14 (2006)
42. McCulloch, W.S., Pitts, W.: A logical calculus of the ideas immanent in nervous activity. Bull. Math. Biophys. **5**(4), 115–133 (1943)
43. Turing, A.M.: I.—Computing machinery and intelligence. Mind **LIX**(236), 433–460 (1950). https://doi.org/10.1093/mind/LIX.236.433.
44. Bradeško, L., Mladenić, D.: A survey of chatbot systems through a loebner prize competition. In: Proceedings of Slovenian Language Technologies Society Eighth Conference of Language Technologies, pp. 34–37 (2012)
45. Samuel, A.L.: Some studies in machine learning using the game of checkers. IBM J. Res. Dev. **3**(3), 210–229 (1959)
46. Newell, A., Shaw, J.C., Simon, H.A.: Report on a general problem solving program. In: IFIP Congress, Pittsburgh, vol. 256, p. 64. (1959)
47. Winograd, T.: Procedures as a representation for data in a computer program for understanding natural language. Tech. rep, Massachusetts Institute of Technology Cambridge Project Mac (1971)
48. Tesauro, G.: Temporal difference learning and TD-Gammon. Commun. ACM **38**(3), 58–68 (1995)
49. Pollack, J.B., Blair, A.D.: Why did TD-gammon work? In: Advances in Neural Information Processing Systems, pp. 10–16 (1997)
50. Silver, D., Huang, A., Maddison, C., Guez, A., Sifre, L., Driessche, G., Schrittwieser, J., Antonoglou, I., Panneershelvam, V., Lanctot, M., Dieleman, S., Grewe, D., Nham, J., Kalchbrenner, N., Sutskever, I., Lillicrap, T., Leach, M., Kavukcuoglu, K., Graepel, T., Hassabis, D.: Mastering the game of go with deep neural networks and tree search. Nature **529**, 484–489 (2016). https://doi.org/10.1038/nature16961
51. Thrun, S., Montemerlo, M., Dahlkamp, H., Stavens, D., Aron, A., Diebel, J., Fong, P., Gale, J., Halpenny, M., Hoffmann, G., Lau, K., Oakley, C., Palatucci, M., Pratt, V., Stang, P., Strohband, S., Dupont, C., Jendrossek, L.E., Koelen, C., Markey, C., Rummel, C., van Niekerk, J., Jensen, E., Alessandrini, P., Bradski, G., Davies, B., Ettinger, S., Kaehler, A., Nefian, A., Mahoney, P.: Winning the DARPA Grand Challenge. J. Field Robot. (2006). Accepted for publication

52. Shotton, J., Fitzgibbon, A., Cook, M., Sharp, T., Finocchio, M., Moore, R., Kipman, A., Blake, A.: Real-time human pose recognition in parts from single depth images. In: CVPR 2011, pp. 1297–1304. IEEE (2011)
53. Ferrucci, D., Brown, E., Chu-Carroll, J., Fan, J., Gondek, D., Kalyanpur, A.A., Lally, A., Murdock, J.W., Nyberg, E., Prager, J., et al.: Building Watson: an overview of the DeepQA project. AI Mag. **31**(3), 59–79 (2010)
54. Hinton, G., Deng, L., Yu, D., Dahl, G.E., Mohamed, A.r., Jaitly, N., Senior, A., Vanhoucke, V., Nguyen, P., Sainath, T.N., et al.: Deep neural networks for acoustic modeling in speech recognition: the shared views of four research groups. IEEE Signal Process. Mag. **29**(6), 82–97 (2012)
55. Silver, D., Hubert, T., Schrittwieser, J., Antonoglou, I., Lai, M., Guez, A., Lanctot, M., Sifre, L., Kumaran, D., Graepel, T., et al.: Mastering chess and shogi by self-play with a general reinforcement learning algorithm. arXiv preprint arXiv:1712.01815 (2017)
56. Heule, M.J.: Schur number five. arXiv preprint arXiv:1711.08076 (2017)
57. Vinyals, O., Babuschkin, I., Chung, J., Mathieu, M., Jaderberg, M., Czarnecki, W., Dudzik, A., Huang, A., Georgiev, P., Powell, R., Ewalds, T., Horgan, D., Kroiss, M., Danihelka, I., Agapiou, J., Oh, J., Dalibard, V., Choi, D., Sifre, L., Sulsky, Y., Vezhnevets, S., Molloy, J., Cai, T., Budden, D., Paine, T., Gulcehre, C., Wang, Z., Pfaff, T., Pohlen, T., Yogatama, D., Cohen, J., McKinney, K., Smith, O., Schaul, T., Lillicrap, T., Apps, C., Kavukcuoglu, K., Hassabis, D., Silver, D.: AlphaStar: Mastering the real-time strategy game StarCraft II. https://deepmind.com/blog/alphastar-mastering-real-time-strategy-game-starcraft-ii/ (2019)
58. Jumper, J., Evans, R., Pritzel, A., Green, T., Figurnov, M., Tunyasuvunakool, K., Ronneberger, O., Bates, R., Žídek, A., Bridgland, A., Meyer, C., Kohl, S.A.A., Potapenko, A., Ballard, A.J., Cowie, A., Romera-Paredes, B., Nikolov, S., Jain, R., Adler, J., Back, T., Petersen, S., Reiman, D., Steinegger, M., Pacholska, M., Silver, D., Vinyals, O., Senior, A.W., Kavukcuoglu, K., Kohli, P., Hassabis, D.: High accuracy protein structure prediction using Deep Learning. https://predictioncenter.org/casp14/doc/CASP14_Abstracts.pdf (2020)
59. Rosenblatt, F.: The perceptron: a probabilistic model for information storage and organization in the brain. Psychol. Review **65**(6), 386 (1958)
60. Minsky, M., Papert, S.: An Introduction to Computational Geometry. Cambridge tiass, HIT (1969)
61. Werbos, P.J.: Applications of advances in nonlinear sensitivity analysis. In: System Modeling and Optimization, pp. 762–770. Springer, Berlin (1982)
62. Kelley, H.J.: Gradient theory of optimal flight paths. ARS J. **30**(10), 947–954 (1960)
63. Platt, J.: Fast training of support vector machines using sequential minimal optimization. In: Advances in Kernel Methods—Support Vector Learning, pp. 185–208. AJ, MIT Press, Cambridge (1999)
64. Rumelhart, D.E., Hinton, G.E., Williams, R.J.: Learning representations by back-propagating errors. Nature **323**(6088), 533–536 (1986)
65. LeCun, Y., Boser, B., Denker, J., Henderson, D., Howard, R., Hubbard, W., Jackel, L.: Handwritten digit recognition with a back-propagation network. Adv. Neural Inf. Process. Syst. **2**, 396–404 (1989)
66. Seewald, AK.: On the Brittleness of handwritten digit recognition Models. ISRN Mach. Vis. **2012** (2012). https://doi.org/10.5402/2012/834127
67. Breiman, L.: Random forests. Mach. Learn. **45**(1), 5–32 (2001)
68. Shamsabadi, A.S., Sanchez-Matilla, R., Cavallaro, A.: Colorfool: Semantic adversarial colorization. In: Proceedings of the IEEE/CVF Conference on Computer Vision and Pattern Recognition, pp. 1151–1160 (2020)
69. Moosavi-Dezfooli, S.M., Fawzi, A., Frossard, P.: Deepfool: a simple and accurate method to fool deep neural networks. In: Proceedings of the IEEE Conference on Computer Vision and Pattern Recognition, pp. 2574–2582 (2016)
70. Kingma, D.P., Ba, J.: Adam: a method for stochastic optimization. arXiv preprint arXiv:1412.6980 (2014)

71. Wikipedia Contributors: comparison of deep-learning software. https://en.wikipedia.org/wiki/Comparison_of_deep-learning_software (2020). Downloaded on 4 Dec 2020
72. Sharma, N., Smeets, B., Tryggestad, C.: The decoupling of GDP and energy growth: a CEO guide. McKinsey Q. (2019)
73. Gordon, R.J.: The rise and fall of American growth: The US standard of living since the civil war, vol. 70. Princeton University Press (2017)
74. Crivellato, E., Ribatti, D.: Soul, mind, brain: Greek philosophy and the birth of neuroscience. Brain Res. Bull. **71**(4), 327–336 (2007). https://doi.org/10.1016/j.brainresbull.2006.09.020
75. Huxley, T.H.: On the hypothesis that animals are automata, and its history. Collected Essays **1** (1874)
76. Wikipedia Contributors: Maillardet's automaton. https://en.wikipedia.org/wiki/Maillardet. Downloaded on 4 Dec 2020
77. Wikipedia contributors: digesting duck. https://en.wikipedia.org/wiki/Digesting_Duck (2020). Downloaded on 4 Dec 2020
78. Wikipedia contributors: Euphonia (Sprachmaschine). https://de.wikipedia.org/wiki/Euphonia_(Sprachmaschine) (2020). Downloaded on 4 Dec 2020
79. Penrose, R.: The emperor's new mind. RSA J. **139**(5420), 506–514 (1991)
80. Adams, B., Petruccione, F.: Quantum effects in the brain: A review. AVS Quant. Sci. **2**(2), 022901 (2020)
81. Kandel, E.R., Schwartz, J.H., Jessell, T.M., of Biochemistry, D., Jessell, M.B.T., Siegelbaum, S., Hudspeth, A.: Principles of neural science, 4th edn. McGraw-Hill, New York (2000)
82. Heindl, A., Schepelmann, M., Nica, R., Ecker, R., Pietschmann, P., Seewald, A.K., Thalhammer, T., Ellinger, I.: Toward the automated detection and characterization of osteoclasts in microscopic images. In: Principles of Osteoimmunology, pp. 31–57. Springer (2016)
83. Seewald, A.K.: Revisiting end-to-end Deep Learning for obstacle avoidance: replication and open issues. In: ICAART, vol. 2, pp. 652–659 (2020)
84. Seewald, A.K., Cypser, J., Mendenhall, A., Johnson, T.: Quantifying phenotypic variation in isogenic Caenorhabditis elegans expressing Phsp-16.2:: gfp by clustering 2D expression patterns. PloS One **5**(7), e11426 (2010)
85. Hofstadter, D.R.: I am a strange loop. Basic Books (2007)
86. Sawyer, R.J.: WWW: wake. Penguin (2009)
87. Hogan, J.P.: The two faces of tomorrow. Baen Books (1997)
88. Hogan, J.P.: Mind matters exploring the world of Artificial Intelligence. Del Ray/Ballantine Publishing Group (1998)
89. Hameroff, S., Penrose, R.: Reply to seven commentaries on "Consciousness in the universe: Review of the 'Orch OR' theory". Phys. Life Rev. **11**(1), 94–100 (2014). https://doi.org/10.1016/j.plrev.2013.11.013

Algorithms, Ethics and Justice

Adnan Hadzi

Abstract This paper discusses the argument that the adoption of artificial intelligence (AI) technologies benefits the powerful few, focussing on their own existential concerns. The paper will narrow down the analysis of the argument to jurisprudence (i.e. the philosophy of law), considering also the historical context. The paper will discuss the construction of the legal system through the lens of political involvement of what one may want to consider to be powerful elites. Before discussing these aspects the paper will clarify the notion of "powerful elites". In doing so the paper will be demonstrating that it is difficult to prove that the adoption of AI technologies is undertaken in a way which mainly serves a powerful class in society. Nevertheless, analysing the culture around AI technologies with regard to the nature of law with a philosophical and sociological focus demonstrates a utilitarian and authoritarian trend in the adoption of AI technologies. The paper will conclude by proposing an alternative, some might say practically unattainable, approach to the current legal system by looking into restorative justice for AI crimes, and how the ethics of care could be applied to AI technologies.

Keywords Power elites · Cyborg · Artificial intelligence · Restorative justice · Legal positivism · Natural law · Disciplinary power · Ethics of care · Privacy

1 Algorithms, Ethics and Justice

In order to lay the foundations for a discussion around the argument that the adoption of artificial intelligence (AI) technologies benefits the powerful few [1, 2], focussing on their own existential concerns [3, 4], the paper will narrow down the analysis of the argument to social justice and jurisprudence (i.e. the philosophy of law), considering also the historical context. This paper signifies an edited version of Hadzi's text on Social Justice and Artificial Intelligence [5], exploring the notion of humanised artificial intelligence [6, 7] in order to discuss potential challenges society

A. Hadzi (✉)
University of Malta, Msida MSD-2080, Malta
e-mail: ahadz01@um.edu.mt

A. Dingli et al. (eds.), *Disruptive Technologies in Media, Arts and Design*, Lecture Notes in Networks and Systems 382, https://doi.org/10.1007/978-3-030-93780-5_9

might face in the future. The paper does not discuss current forms and applications of artificial intelligence, as, so far, there is no AI technology [8], which is self-conscious and self-aware, being able to deal with emotional and social intelligence. It is a discussion around AI as a speculative hypothetical entity. One could then ask, if such a speculative self-conscious hardware/software system were created at what point could one talk of personhood? And what criteria could there be in order to say an AI system was capable of committing AI crimes?

In order to address AI crimes, the paper will start by outlining what might constitute personhood in discussing legal positivism and natural law. Concerning what constitutes AI crimes the paper uses the criteria given in King et al.'s paper Artificial Intelligence Crime: An Interdisciplinary Analysis of Foreseeable Threats and Solutions [9], where King et al. coin the term AI crime, mapping five areas in which AI might, in the foreseeable future, commit crimes, namely:

- commerce, financial markets, and insolvency
- harmful or dangerous drugs
- offences against persons
- sexual offences
- theft and fraud, and forgery and personation

Having those potential AI crimes in mind, the paper will discuss the construction of the legal system through the lens of political involvement of what one may want to consider to be powerful elites. Before discussing these aspects the paper will clarify the notion of "powerful elites". In doing so the paper will be demonstrating that it is difficult to prove that the adoption of AI technologies is undertaken in a way which mainly serves a powerful class in society. Nevertheless, analysing the culture around AI technologies with regard to the nature of law with a philosophical and sociological focus enables one to demonstrate a utilitarian and authoritarian trend in the adoption of AI technologies [10–13].

The paper will base the discussion around Crook's notion on "power elites" [14], in Media Law and Ethics [15], and apply it to the discourse around artificial Intelligence and ethics. Following Crook the paper will introduce a discussion around power elites with the notions of legal positivism and natural law, as discussed in the academic fields of philosophy and sociology. The paper will then look, in a more detailed manner, into theories analysing the historical and social systematisation, or one may say disposition, of laws, and the impingement of neo-liberal [16] tendencies upon the adoption of AI technologies. Pueyo demonstrates those tendencies with a thought experiment around superintelligence in a neoliberal scenario [17]. In Puevo's thought experiment the system becomes techno-social-psychological with the progressive incorporation of decision-making algorithms and the increasing opacity of such algorithms [18], with human thinking partly shaped by firms themselves [19].

The regulatory, self-governing potential of AI algorithms [20–22] and the justification by authority of the current adoption of AI technologies within civil society will be analysed next. The paper will propose an alternative, some might say practically unattainable, approach to the current legal system by looking into restorative justice for AI crimes [23], and how the ethics of care, through social contracts, could be applied to AI technologies. In conclusion the paper will discuss affect [24, 25] and humanised artificial intelligence with regards to the emotion of shame, when dealing with AI crimes.

2 Legal Positivism and Natural Law

In order to discuss AI in relation to personhood this paper follows the descriptive psychology method [26] of the paradigm case formulation [27] developed by Ossorio [28]. Similar to how some animal rights activists call [29] for certain animals to be recognised as non-human persons [30], this paper speculates on the notion of AI as a non-human person being able to reflect on ethical concerns [31, 32]. Here Schwartz argues that "it is reasonable to include non-humans as persons and to have legitimate grounds for disagreeing where the line is properly drawn. In good faith, competent judges using this formulation can clearly point to where and why they agree or disagree on what is to be included in the category of persons" [33].

According to Ossorio [26] a deliberate action is a form of behaviour in which a person (a) engages in an intentional action, (b) is cognizant of that, and (c) has chosen to do that. Ossorio gives four classifications: ethical, hedonic, aesthetic, and prudent as fundamental motivations. Ethical motivations, as well as aesthetic motivations, can be distinguished from prudent (and hedonic) motivations due to the agent making a choice. "Aesthetic and ethical motivations are only relevant when deliberate action is also possible since aesthetic and ethical action require the eligibility to choose or refrain, to potentially deliberate about the desirable course to follow. In the service of being able to choose, and perhaps think through the available options, a person's aesthetic and ethical motives are often consciously available" [34].

In the fields of philosophy and sociology countless theories have been advanced concerning the nature of law, addressing questions such as: Can unethical law be binding? Should there be a legal code for civil society? Can such a legal code be equitable, unbiased, and just, or, is the legal code always biased? In the case of AI technologies one can ask whether the current vision for the adoption of AI technologies is a vision that benefits only the powerful elites.

To address the question one needs to discuss the idea of equality. Reference is made to Aristotle's account on how the legal code should be enacted in an unbiased manner [35]. Aristotle differentiated between an unbalanced and balanced application of the legal code, pointing out that the balanced juridical discussion of a case should be courteous. Here, as with the above mentioned animal rights activists, in Dependent Rational Animals MacIntyre [36] argued, drawing on Aquina's [37] discussion of misericordia, for the recognition of our kinship to some species calling for the "virtues of acknowledged dependence" [38]. Austin, on the other hand, suggests that the legal code is defined by a higher power, "God", to establish justice over society. For Austin the legal code is an obligation, a mandate to control society [39].

Hart goes on to discuss the social aspect of legal code and how society apprehends the enactment of such legal code [40]. Hart argues that the legal code is a strategy, a manipulation of standards accepted by society. Contrary to Hart, Dworkin proposes for the legal code to allow for non-rule [41] standards reflecting ethical conventions of society. Dworkin discusses legislation as an assimilation of these conventions, where legislators do not define the legal code, but analyse the already existing conventions to derive conclusions, which then in turn define the legal code. Nevertheless, Dworkin fails to explain how those conventions come into being. Here for Kelsen [42, 43] legal code is a product of the political, cultural and historical circumstances society finds itself in. For Kelsen the legal code is a standardising arrangement which defines how society should operate [44].

The paradigm case [26] allows for the potential AI as non-human persons [45, 46]. Referring to the paradigm case method allows one to work out where parties are in agreement or disagreement concerning what constitutes a person.

Here social contract theories, as defined and discussed below, might serve to explain and analyse how legal codes deal with the emergence of legal issues concerning AI technologies or AI crimes. Following Ossorio [28] since persons act consciously, they are motivated by ethical, aesthetic, prudent and hedonic motivations: at the same time, social contract(s) allowing persons to act in patterns of significance, giving meaning to one's actions.

AI can be interpreted as automated distribution systems, using data drawn from a 'datasphere', which could easily be imagined continuously operating without human interference. Thus, a more particular definition of 'datasphere' would emphasise how a vast amount of data circulates, while only becoming meaningful when viewed in the context of a social contract. In other words, the transformation of 'data' into 'meaning' can always be seen to take place within a social contract. For example, a protocol extracting data always has to be configured, i.e. socially or politically agreed upon. Legal or activist interventions thus always interpellate the datasphere. dataspheres include all forms of data that exist in the public domain and public spheres. This data becomes meaningful only when actors interpret it. Such instances of interaction are always in some ways social.

In that sense a legal system, social contracts, aiming to control the dataspheres, needs to be tailored carefully because the situation as being controlled by the most driven producers and consumers. The old distribution model is so impoverished that it chooses the safest route. Applying the notion of 'social contracts', the notion of open and distributed sharing can be reinforced as an overall heuristic and social ethos. One can even elaborate upon the idea of slavery, extending it to the idea of social contracts with reference to Jean-Jacques Rousseau's Social Contract, which states: "The words 'slavery' and 'right' are contradictory, they cancel each other out. Whether as between one man and another, or between one man and a whole people, it would always be absurd to say: I hereby make a covenant with you which is wholly at your expense and wholly to my advantage" [47, p. 58].

"Man is born free; and everywhere he is in chains", begins Rousseau's work of political philosophy, The Social Contract [47]. Rousseau [48, 49] aimed to understand why "a man would give up his natural freedoms and bind himself to the rule of a prince or a government" [50]. This question of political philosophy was widely discussed in the seventeenth and eighteenth centuries, as revolution was in the air all over Europe, particularly in France 1789. In the eighteenth century Rousseau published The Social Contract. Rousseau thought that there is a conflict between obedience and persons' freedom and argued that our natural freedom is our own will. Rousseau defined the social contract as a law 'written' by everybody [51]. His argument was that if everybody was involved in making the laws they would only have to obey themselves and as such follow their free will. How could persons then create a common will? For Rousseau this would only have been possible in smaller communities through the practice of caring for each other and managing conflicts for the common good—ultimately through love. In The Art of Loving Erich Fromm reminds us that "love is not a sentiment which can be easily indulged in by anyone … [S]atisfaction in individual love cannot be attained without the capacity to love one's neighbour, without true humility, courage, faith and discipline" [52, p. xix]. Rousseau imagined a society the size of his native city of Geneva as an ideal ground for the implementation of social contract theory. Ironically it was the French who, through their revolutionaries, implemented social contract theory. Nevertheless, the French people read it differently, as imposing social contracts onto the persons. The mass-scale imposition of contracts compromised their non-mandatory status.

In the twentieth century, moral and political theory around the social contract had a revival with John Rawls' A Theory of Justice [53] and Gauthier's Morals by Agreement [54]. Gauthier argues after Hobbes [55] and explains that there can be morality in our society without the state having to impose morality with the help of external enforcement mechanisms. For Gauthier rationality is the key for cooperation and for following agreements made between different parties. Celeste Friend states in Social Contract Theory [56] that feminist philosophers criticise social contract theory for not reflecting moral and political lives correctly and completely, and for the contract itself being "parasitical upon the subjugations of classes of persons" [56].

In a more critical approach to rationalized contracts, in The Sexual Contract Carole Pateman argues that "lying beneath the myth of the idealized contract, as described by Hobbes, Locke, and Rousseau, is a more fundamental contract concerning men's relationship to women" [56]. Similarly, for Pateman, "[t]he story of the sexual contract reveals that there is good reason why 'the prostitute' is a female figure" [57, p. 192]. The feminist philosophers Baier [58, 59] and Held [60, 61] criticise social contract theory for not demonstrating fully what a moral person should be and how this affects relationships. Baier argues that Gauthier does not reflect on the full spectrum of human motivations and their psychology, that he fails to see that there is a dependency on certain relationships (like mother–child) before one can enter into those contracts, as described in Baier's expression "the cost of free milk" [58]. Held, as quoted by Friend, even goes so far as to argue that "contemporary Western society is in the grip of contractual thinking" [56].

In The Racial Contract, Mills [62] inspired by The Sexual Contract argues that non-whites have similar problems with the class society as women, both sets of conflicts and suppression deriving from a patriarchal mindset. For Mills there is a 'racial contract' which is more important to the industrialized part of the world than the social contract, which one might want to consider in relation to humanised artificial intelligent systems. "This racial contract determines in the first place who counts as fully moral and political persons, and therefore sets the parameters of who can 'contract in' to the freedom and equality that the social contract promises" [56].

The subject of the Debian Social Contract [63] might very well be the one who writes most of the code for the data sphere and defines AI technologies: the white male [64]. Taking the above criticism regarding the sexual and the racial contract on board one could extend the discussion on social contracts with the notion of Open Contracts. First one needs to look into the current Debian Social Contract and the issue of privacy with regard to Intellectual Property [65]. The Debian Foundation is one of the biggest communities for the Linux [66] operating system. The beginning of the Debian Social Contract for the FLOSS community states:

"Our priorities are our users and free software. We will be guided by the needs of our users and the free software community. We will place their interests first in our priorities. We will support the needs of our users for operation in many different kinds of computing environments. We will not object to non-free works that are intended to be used on Debian systems, or attempt to charge a fee to people who create or use such works. We will allow others to create distributions containing both the Debian system and other works, without any fee from us. In furtherance of these goals, we will provide an integrated system of high-quality materials with no legal restrictions that would prevent such uses of the system" [63].

The idea of the Debian Social Contract could be extended to AI technologies, in the form of Open Contracts, suggesting similar principles that can be applied to free and open source software. One can argue that these would be a pre-condition for 'ethical' AI technologies. With open contracts such as the Debian Social Contract in place, various communities can start discussing, experimenting with and practising the production, distribution, and sharing of AI technologies. Although this sounds like a promising scenario one also has to be critical, as these alternatives can be vulnerable to corruption.

One could support an Open Contract practice, and suggest that a feminist notion of 'restorative justice' [15, 67] might serve to judge Open Contracts, by applying the notions of solidarity and care as principles of judicial practice. However the concern is how to move from an abstract idea of open contracts to a concrete legislation which could enable a AI technology production that is not deemed antithetical, or oppositional to the current judicial system, by formulating a set of ground rules and protocols that will allow AI communities to function and prosper. One could argue that this can be done by defining the independent terms and conditions, namely free and open licenses. Social contracts and laws will eventually be defined for these dataspheres, but until then power elites will try to appropriate every piece of AI technology in accord with the old, non-efficacious, "IP legislation" [68].

Nevertheless, in trying to evaluate the argument that the adoption of AI technologies is a process controlled by powerful elites who wield the law to their benefit, one also needs to discuss the notion of power elites. Chamblis and Seidman argue that powerful interests have shaped the writing of legal codes for a long time [69]. However, Chamblis and Seidman also state that legislation derives from a variety of interests, which are often in conflict with each other. One needs to extend the analysis not only to powerful elites, but one also needs to examine the notion of power itself, and the extent to which power shapes legislation, or, on the contrary, if it is legislation itself that controls power.

In an attempt to identify the source of legislation, Weber argues that legal code is powerfully interlinked with the economy. Weber goes on to argue that this link is the basis of capitalist society [70]. Here one can refer back to Marx's idea of materialism and the influence of class society on legislation [71]. For Marx legislation, legal code, is an outcome of the capitalist mode of production [72]. Marx's ideas have been widely discussed with regards to the ideology behind the legal code. Nevertheless Marx's argumentation limits legal code to the notion of class domination.

Here Sumner extended on Marx's theories regarding legislation and ideology and discussed the legal code as an outcome of political and cultural discussions, based on the economic class domination [73]. Sumner expands the conception of the legal code not only as a product of the ruling class but also as bearing the imprint of other classes, including blue-collar workers, through culture and politics. Sumner argues that with the emergence of capitalist society, "the social relations of legal practice were transformed into commercial relations" (ibid: 51). However, Sumner does not discuss why parts of society are sidelined by legislation, and how capitalist society not only impacts on legislation, but also has its roots in the neo-liberal writing of legal code.

To apprehend how ownership, property and intellectual rights became enshrined in legal code and adapted by society one can turn to Locke's theories [74]. Locke argued that politicians ought to look after ownership rights and to support circumstances allowing for the growth of wealth (capital). Following Locke one can conclude that contemporary society is one in which politicians influence legislation in the interest of a powerful upper-class—a neo-liberal society. Still, one needs to ask, should this be the case, and should powerful elites have the authority over legal code, how legislation is enacted and maintained?

3 The Disciplinary Power of Artificial Intelligence

In order to discuss these questions one has to analyse the history of AI technologies leading to the kind of "humanised" AI system this paper posits. Already in the 50s Turing, the inventor of the Turing test [75], had stated that:

"We may hope that machines will eventually compete with men in all purely intellectual fields. But which are the best ones to start with? Even this is a difficult decision. Many people think that a very abstract activity, like the playing of chess, would be best. It can also be maintained that it is best to provide the machine with the best sense organs that money can buy, and then teach it to understand and speak English. This process could follow the normal teaching of a child. Things would be pointed out and named, etc. Again I do not know what the right answer is, but I think both approaches should be tried. We can only see a short distance ahead, but we can see plenty there that needs to be done" [76].

The old fashioned approach [77], some may say still contemporary approach, was to primarily research into 'mind-only' [78] AI technologies/systems. Through high level reasoning, researchers were optimistic that AI technology would quickly become a reality.

Those early AI technologies were a disembodied approach using high level logical and abstract symbols. By the end of the 80s researchers found that the disembodied approach was not even achieving low level tasks humans could easily perform [79]. During that period many researchers stopped working on AI technologies and systems, and the period is often referred to as the 'AI winter' [80, 81].

Brooks then came forward with the proposition of 'Nouvelle AI' [82], arguing that the old fashioned approach did not take into consideration motor skills and neural networks. Only by the end of the 90s did researchers develop statistical AI [79] systems without the need for any high level logical reasoning; instead AI systems were 'guessing' through algorithms and machine learning. This signalled a first step towards humanistic artificial intelligence, as this resembles how humans make intuitive decisions [83] here researchers suggest that embodiment improves cognition [84, 85].

With embodiment theory Brooks argued that AI systems would operate best when computing only the data that was absolutely necessary [86]. Further in Developing Embodied Multisensory Dialogue Agents Paradowski [87] argues that without considering embodiment, e.g. the physics of the brain, it is not possible to create AI technologies/systems capable of comprehension, and that AI technology "could benefit from strengthened associative connections in the optimization of their processes and their reactivity and sensitivity to environmental stimuli, and in situated human–machine interaction. The concept of multisensory integration should be extended to cover linguistic input and the complementary information combined from temporally coincident sensory impressions" [87].

With this historical analysis in mind one can discuss the paper's focus on power elites. Raz studied the procedures through which elites attain disciplinary power in society [88]. Raz argues that the notion of the disciplinary power of elites in society is exchangeable with the disciplinary power of legislation and legal code. Raz explains that legal code is perceived by society as the custodian of public order. He further explains that by precluding objectionable actions, legislation directs society's activities in a manner appropriate to jurisprudence. Nevertheless, Raz did not demonstrate how legislation impacts on personal actions. This is where Foucault's theories on discipline and power come in. According to Foucault the disciplinary power of legislation leads to a self-discipline of individuals [89]. Foucault argues that the institutions of courts and judges motivate such a self-disciplining of individuals [90], and that self-disciplining rules serve "more and more as a norm" [91].

Foucault's theories are especially helpful in discussing how the "rule of truth" has disciplined civilisation and how power elites, as institutions, push through an adoption of AI technologies which seem to benefit mainly the upper-class. Discussions around truth, Foucault states, form legislation into something that "decides, transmits and itself extends upon the effects of power" [92, p. 230]. Foucault's theories help to explain how legislation, as an institution, is rolled out throughout society with very little resistance, or "proletarian counter-justice" [93, p. 34]. Foucault explains that this has made the justice system and legislation a for-profit system. With this understanding of legislation, and social justice, one does need to reflect further on Foucault's notion of how disciplinary power seeks to express its distributed nature in the modern state. Namely one has to analyse the distributed nature of those AI technologies, especially through networks and protocols, so that the link can now be made to AI technologies becoming "legally" more profitable, in the hands of the upper-class.

If power generates new opportunities rather than simply repressing them, then, following Foucault [94], more interaction and participation can extend and not simply challenge power relations. Foucault's text The Subject and Power [95] offers a valuable insight into power relationships relevant also within AI technologies. It is the product of research that was undertaken by Foucault over a period of over twenty years. Foucault uses the metaphor of a chemical catalyst for a resistance which can bring to light power relationships, and thus allow an analysis of the methods this power uses: "[r]ather than analysing power from the point of view of its internal rationality, it consists of analysing power relations through the antagonism of strategies" [95, p. 780].

In Protocol, Galloway describes how these protocols changed the notion of power and how "control exists after decentralization" [96, p. 81]. Galloway argues that protocol has a close connection to both Deleuze's concept of 'control' and Foucault's concept of biopolitics [97, pp. 1978–1979] by claiming that the key to perceiving protocol as power is to acknowledge that "protocol is an affective, aesthetic force that has control over life itself" [96, p. 81]. Galloway suggests [96, p. 147] that it is important to discuss more than the technologies, and to look into the structures of control within technological systems, which also include underlying codes and protocols, in order to distinguish between methods that can support collective production, e.g. sharing of AI technologies within society, and those that put the AI technologies in the hands of the powerful few. Galloway's argument in the chapter Hacking [96, p. 146] is that the existence of protocols "not only installs control into a terrain that on its surface appears actively to resist it", but goes on to create the highly controlled network environment. For Galloway hacking is "an index of protocological transformations taking place in the broader world of techno-culture" [96, p. 157].

In order to be able to regulate networks and AI technologies, control and censorship mechanisms are introduced to networks by applying them to devices and nodes. This form of surveillance, or dataveillance, might constitute a development akin to Michel Foucault's concept of "panopticism" [98], "panoptic apparatus" [99, p. 5], defined as both massive collections and storage of vast quantities of personal data and the systemic use of such data in the investigation or monitoring of one or more persons. Laws and agreements like the Anti-Counterfeiting Trade Agreement [100, 101], the Digital Economy Act and the Digital Millennium Copyright Act require surveillance of the AI technologies that consumers use in their "private spheres" [102–104], and can be used to silence "critical voices" [105]. The censorship of truth, and the creation of fear of law through moral panics stand in opposition to the development of a healthy democratic use of AI technologies. Issues regarding the ethics of AI [106–109] arise from this debate.

Fitzpatrick expands on Foucault's theory, investigating the "symbiotic link between the rule of law and modern administration" [110]. Fitzpatrick states that legal code is not only a consequence of disciplinary power, but that it also legalises dubious scientific experiments. Here again one can make the link to ethical questionable advances with AI technologies. Legislation, or legal code, Fizpatrick argues, corrects "the disturbance of things in their course and reassert the nature of things" (ibid, p. 160). For Fitzpatrick legislation is not an all-embracing, comprehensive concept as argued by Dworkin [41] and Hart [39], but rather legislation is defined by elites. For Fitzpatrick legislation "changes as society changes and it can even disappear when the social conditions that created it disappear or when they change into conditions antithetical to it" [110, p. 6]. Furthermore, West [111] suggests that the impact of disciplinary power through legislation on the belief system of individuals does not allow for an analytical, critical engagement by individuals with the issues at stake. Legislation is simply regarded as given. In relation to the disciplinary power of AI technologies, issues with privacy, defamation and intellectual property laws are not being questioned. Nevertheless, West's argument that all individuals adhere to equivalent morals is improbable.

Adams and Brownsword [112] give a more nuanced view of contemporary legislation. They argue that legislation aims to institute public order. Legislation sets up authoritative mechanisms whereby social order can be established and maintained, social change managed, disputes settled and policies and goals for the community adopted (ibid: 11). Adams and Brownsword go on to argue that legal code is skewed in favour of the upper-class and those who engage more with politics in society—examples of which could be the corporate sector producing AI technologies and business elites seeking to use AI technologies for profit. According to Adams and Brownsword there seems to be no unbiased, fair legislation or legal code, and the maintenance of public order must simply reproduce an unfair class society. If this is the case, following Adams and Brownsword argumentation, one can argue that indeed the adoption of AI technologies does not follow a utilitarian ethical code, benefiting society, but rather conforms to the interests of a small group, those owning AI technologies.

A further discussion of disciplinary power within the process of writing legal code is that of Chamblis and Seidman [69], who argue that legislation is not produced through a process characterised by balanced, fair development, but rather by powerful elites writing legal code by themselves. Translating this again back to the adoption of AI technologies, it becomes evident that the freedom to engage with those technologies is left to those who have the financial means, and with it the legal means, to do so. According to Chamblis and Seidman, in a culture dominated by economics, legislation and technologies are being outlined and modelled by those powerful elites.

The analysis of the theories above has attempted to show that the implementation of AI technologies might be construed as a project deriving from, and serving the interests of, the dominant class; following Foucault's terminology, this is achieved using the disciplinary power of legislation, through regimes of truths, over individuals.

AI technologies, rather than benefiting society, could very well be implemented against society. The implementation of AI technologies follows legislation set out by elites, raising issues connected with privacy, national security, or intellectual property laws. On this note, Crook states that "there is the risk that their decisions are based on profit and loss rather than truth/justice and freedom of expression" [15, p. 94].

4 AI Technologies and Restorative Justice: The Ethics of Care

Having said this, the prospect could be raised that restorative justice might offer "a solution that could deliver more meaningful justice" [15, p. 310]. With respect to AI technologies, and the potential inherent in them for AI crimes, instead of following a retributive legislative approach, an ethical discourse [113], with a deeper consideration for the sufferers of AI crimes [114] should be adopted. That said, acting ethically is more difficult than ever [115], due to the hyper expansion of big data and artificial intelligence [116, 117]. Research into artificial intelligence has gone from being a public service undertaken mainly at universities to being run (and regarded) as businesses, run by big corporations such as Alphabet (parent company of Google) and Facebook, created to generate profit [118]. The companies need to attract a large number of paying customers. AI technologies have become workers in the market economy, rarely following any ethical guidelines [119]. One can ask: could restorative justice offer an alternative way of dealing with the occurrence of AI crimes [120, 121]?

Vidmar and Millar described two psychological perceptions of justice [122]. One is behavioural control, following the legal code as strictly as possible, punishing any wrongdoer [123], and second the restorative justice system, which focuses on restoration where harm was done. Thus an alternative approach for the ethical implementation of AI technologies, with respect to legislation, might be to follow restorative justice principles. Restorative justice would allow for AI technologies to learn how to care about ethics [124, 125]. Fionda [126] describes restorative justice as a conciliation between victim and offender, during which the offence is deliberated upon. Both parties try to come to an agreement on how to achieve restoration for the damage done, to the situation before the crime (here an AI crime) happened. Restorative justice advocates compassion for the victim and offender, and a consciousness on the part of the offenders as to the repercussion of their crimes. Tocqueville argued for one to live in liberty, "it is necessary to submit to the inevitable evils which it engenders" [127].

One can argue that these evils are becoming more evident nowadays with the advance of AI technologies. For AI crimes punishment in the classical sense may seem to be adequate [128]. Duff [129] argues that using a punitive approach to punish offences educates the public. Wenzel and Okimoto [123] refer to Durkheim's studies on the social function of punishment [130], serving to establish a societal awareness of what ought to be right or wrong. Christie [67], however, criticises this form of execution of the law. He argues that, through conflict, there is the potential to discuss the rules given by law, allowing for a restorative process, rather than a process characterised by punishment and a strict following of rules. Christie states that those suffering most from crimes are suffering twice, as although it is the offenders being put on trial, the victims have very little say in courtroom hearings where mainly lawyers argue with one-another. It basically boils down to guilty or not guilty, and no discussion in between. Christie argues that running restorative conferencing sessions helps both sides to come to terms with what happened. The victims of AI crimes would not only be placed in front of a court, but also be offered engagement in the process of seeking justice and restoration.

Restorative justice might support victims of AI crimes better than the punitive legal system, as it allows for the sufferers of AI crimes to be heard in a personalised way, which could be adopted to the needs of the victims (and offenders). As victims and offenders represent themselves in restorative conferencing sessions, these become much more affordable [131], meaning that the barrier to seeking justice due to the financial costs would be partly eliminated, allowing for poor parties to be able to contribute to the process of justice. This would benefit wider society and AI technologies would not only be defined by a powerful elite. Restorative justice could hold the potential not only to discuss the AI crimes themselves, but also to get to the root of the problem and discuss the cause of an AI crime. For Braithwaite [132] restorative justice makes re-offending harder.

In such a scenario, a future AI system capable of committing AI crimes would need to have a knowledge of ethics around the particular discourse of restorative justice. The implementation of AI technologies will lead to a discourse [133] around who is responsible for actions taken by AI technologies. Even when considering clearly defined ethical guidelines, these might be difficult to implement [134] due to the pressure of competition AI systems find themselves in.

That said, this speculation is restricted to humanised artificial intelligence systems to be part of a restorative justice system, through the very human emotion of shame. Without a clear understanding of shame [135] it will be impossible to resolve AI crimes in a restorative manner. Thus one might want to think about a humanised, cyborgian [136, 137] proposal of a symbiosis between humans and technology, along the lines of Kasparov's advanced chess [138], as in advanced jurisprudence [139], a legal system where human and machine work together on restoring justice, for social justice.

References

1. Chaslot, G.: YouTube's A.I. was divisive in the US presidential election. Medium, Nov. 27, 2016. https://medium.com/the-graph/youtubes-ai-is-neutral-towards-clicks-but-is-biased-towards-people-and-ideas-3a2f643dea9a#.tjuusil7d. Accessed 25 Feb 2018

2. Morozov, E.: London. The geopolitics of artificial intelligence (2018). [Online Video]. Available: https://www.youtube.com/watch?v=7g0hx9LPBq8. Accessed 11 July 2018
3. Busby, M.: Use of "killer robots" in wars would breach law, say campaigners. The Guardian, Aug. 21, 2018. [Online]. Available: https://web.archive.org/web/20181203074423/https://www.theguardian.com/science/2018/aug/21/use-of-killer-robots-in-wars-would-breach-law-say-campaigners. Accessed 21 Aug 2018
4. Sample, I.: Thousands of scientists pledge not to help build killer AI robots. The Guardian, July 18, 2018. [Online]. Available: http://www.theguardian.com/science/2018/jul/18/thousands-of-scientists-pledge-not-to-help-build-killer-ai-robots. Accessed 18 July 2018
5. Hadzi, A.: Social justice and artificial intelligence. Body Space Technol. **18**(1), Art. no. 1 (2019). https://doi.org/10.16995/bst.318
6. Kaplan, A., Haenlein, M.: Siri, Siri, in my hand: who's the fairest in the land? On the interpretations, illustrations, and implications of artificial intelligence. Bus. Horiz. **62**(1), 15–25 (2019). https://doi.org/10.1016/j.bushor.2018.08.004
7. Legg, S., Hutter, M.: A Collection of Definitions of Intelligence. IDSIA, Lugano, Switzerland (2007). [Online]. Available: http://arxiv.org/abs/0706.3639. Accessed 4 Jan 2019
8. Bostrom, N.: Superintelligence: Paths, Dangers, Strategies. Oxford University Press, Oxford (2014)
9. King, T., Aggarwal, N., Taddeo, M., Floridi, L.: Artificial intelligence crime: an interdisciplinary analysis of foreseeable threats and solutions. Social Science Research Network, Rochester, NY, SSRN Scholarly Paper ID 3183238, May 2018. [Online]. Available: https://papers.ssrn.com/abstract=3183238. Accessed 4 Jan 2019
10. Goodman, J.: Robots in Law: How Artificial Intelligence is Transforming Legal Services. ARK Group (2016)
11. Haddadin, S.: Towards Safe Robots: Approaching Asimov's 1st Law. Springer, London (2013)
12. Hallevy, G.: When Robots Kill: Artificial Intelligence Under Criminal Law. UPNE, London (2013)
13. Pagallo, U.: The Laws of Robots: Crimes, Contracts, and Torts. Springer Science & Business Media (2013)
14. Crook, T.: Power, intelligence, whistle-blowing and the contingency of history. Presented at the Annual Conference of the Institute of Communication Ethics, The Foreign Press Association, London (2010). [Online]. Available: https://www.gold.ac.uk/media-communications/staff/crook/. Accessed 30 Sept 2018
15. Crook, T.: Comparative Media Law and Ethics. Routledge, London (2009)
16. Parikh, P.: On liberalism and neoliberalism. Medium, Oct 21, 2017. https://medium.com/@pparikh1/on-liberalism-and-neoliberalism-5946523aa2ca. Accessed 4 Jan 2019
17. Pueyo, S.: Growth, degrowth, and the challenge of artificial superintelligence. J. Clean. Prod. **197**, 1731–1736 (2018). https://doi.org/10.1016/j.jclepro.2016.12.138
18. Danaher, J.: The threat of algocracy: reality, resistance and accommodation. Philos. Technol. **29**(3), 245–268 (2016). https://doi.org/10.1007/s13347-015-0211-1
19. Galbraith, J.K.: The New Industrial State. Princeton University Press, Oxford (2015)
20. Poole, S.: Arabic, algae and AI: the truth about "algorithms". The Guardian, Sept 20, 2018. [Online]. Available: https://web.archive.org/web/20181119100303/https://www.theguardian.com/books/2018/sep/20/from-arabic-to-algae-like-ai-the-alarming-rise-of-the-algorithm. Accessed 20 Sept 2018
21. Roio, D.: Algorithmic sovereignty, Thesis, University of Plymouth (2018). [Online]. Available: https://pearl.plymouth.ac.uk/handle/10026.1/11101. Accessed 20 March 2018
22. Smith, A.: Franken-algorithms: the deadly consequences of unpredictable code. The Guardian, Aug 30, 2018. [Online]. Available: https://web.archive.org/web/20190105054549/https://www.theguardian.com/technology/2018/aug/29/coding-algorithms-frankenalgos-program-danger. Accessed 2 Sept 2018
23. Cadwalladr, C.: Elizabeth Denham: "data crimes are real crimes". The Guardian, July 15, 2018. [Online]. Available: https://web.archive.org/web/20181121235057/https://www.theguardian.com/uk-news/2018/jul/15/elizabeth-denham-data-protection-information-commissioner-facebook-cambridge-analytica. Accessed 18 July 2018

24. Olivier, B.: Cyberspace, simulation, artificial intelligence, affectionate machines and being human. Communicatio **38**(3), 261–278 (2012). https://doi.org/10.1080/02500167.2012.716763
25. Wilson, E.A.: Affect and Artificial Intelligence. University of Washington Press, Washington (2011)
26. Ossorio, P.G.: The Behavior of Persons. Descriptive Psychology Press, Ann Arbor Mich. (2013). [Online]. Available: http://www.sdp.org/sdppubs-publications/the-behavior-of-persons/
27. Jeffrey, J.: Knowledge engineering: theory and practice. Soc. Descr. Psychol. **5**, 105–122 (1990)
28. Ossorio, P.G.: Persons: The Collected Works of Peter G. Ossorio, vol. I, Descriptive Psychology Press, Ann Arbor Mich. (1995). [Online]. Available: http://www.sdp.org/sdppubs-publications/persons-the-collected-works-of-peter-g-ossorio-volume-1/
29. Mountain, M.: Lawsuit filed today on behalf of chimpanzee seeking legal personhood, Nonhuman Rights Project, Dec 2, 2013. https://www.nonhumanrights.org/blog/lawsuit-filed-today-on-behalf-of-chimpanzee-seeking-legal-personhood/. Accessed 8 Jan 2019
30. Midgley, M.: Fellow champions dolphins as "non-human persons", Oxford Centre for Animal Ethics (2010). https://www.oxfordanimalethics.com/2010/01/fellow-champions-dolphins-as-%E2%80%9Cnon-human-persons%E2%80%9D/. Accessed 8 Jan 2019
31. Bergner, R.: The Tolstoy dilemma: a paradigm case formulation and some therapeutic interventions. Adv. Descr. Psychol. **9** (2010). [Online]. Available: http://www.sdp.org/sdppubs-publications/advances-in-descriptive-psychology-vol-9/. Accessed 8 Jan 2019
32. Laungani, P.: Mindless psychiatry and dubious ethics. Couns. Psychol. Q. **15**(1), 23–33 (2002). https://doi.org/10.1080/09515070110102305
33. Schwartz, W.: What is a person and how can we be sure? A paradigm case formulation. Social Science Research Network, Rochester, NY, SSRN Scholarly Paper ID 2511486, 2014. [Online]. Available: https://papers.ssrn.com/abstract=2511486. Accessed 8 Jan 2019
34. Schwartz, W.: The two concepts of action and responsibility in psychoanalysis. J. Am. Psychoanal. Assoc. **32**(3), 557–572 (1984). https://doi.org/10.1177/000306518403200306
35. Aristotle, Saunders, T.J.: The Politics. Penguin UK, London (1981)
36. MacIntyre, A.: Dependent Rational Animals: Why Human Beings Need the Virtues, Revised Open Court, Chicago (2001)
37. Aquinas, T.: Summa Theologiae, vol. 33, Hope: 2a2ae. 17–22. Cambridge University Press, Cambridge 2006
38. MacIntyre, A.: After Virtue. A&C Black, London (2013)
39. Austin, J.: The Province of Jurisprudence Determined: And, the Uses of the Study of Jurisprudence. Hackett Publishing (1998)
40. Hart, H.L.A.: The Concept of Law. Oxford University Press, Oxford (1961)
41. Dworkin, R.: A Matter of Principle. Clarendon Press, Oxford (1986)
42. Kelsen, H.: Pure Theory of Law. University of California Press, Los Angeles (1967)
43. Kelsen, H.: General Theory of Law and State. The Lawbook Exchange Ltd., New Jersey (2009)
44. Kelsen, H.: General Theory of Norms. Clarendon Press, Oxford (1991)
45. Putman, A.: Artificial persons. Adv. Descr. Psychol. **5**, 81–104 (1990)
46. Schwartz, W.: The problem of other possible persons: dolphins, primates and aliens. Social Science Research Network, Rochester, NY, SSRN Scholarly Paper ID 2402230, 1982. [Online]. Available: https://papers.ssrn.com/abstract=2402230. Accessed 8 Jan 2019
47. Rousseau, J.J.: The Social Contract. Penguin Classics, London 1968. [Online]. Available: http://www.constitution.org/jjr/socon.htm. Accessed 20 May 2010
48. Dart, G.: Rousseau, Robespierre and English Romanticism. Cambridge University Press, Cambridge (2005)
49. Hampsher-Monk, I.: A History of Modern Political Thought. Wiley-Blackwell, New York (1992)

50. Bragg, M.: The Social Contract. In our Time, BBC 4, London, Feb 7, 2008. [Online]. Available: http://www.bbc.co.uk/radio4/history/inourtime/inourtime_20080207.shtml. Accessed 31 March 2009
51. Roland, J.: The Social Contract and Constitutional Republics. Constitution.org (1994). http://www.constitution.org/soclcont.htm. Accessed 19 Oct 2009
52. Fromm, E.: The Art of Loving. Harper & Row, New York (1956)
53. Rawls, J.: A Theory of Justice. Harvard University Press, Cambridge, MA (2005)
54. Gauthier, D.: Morals by Agreement. Clarendon Press, Alderley (1986)
55. Hobbes, T.: Leviathan. London: Andrew and William Crooke. St. Paul's Churchyard, 1651. [Online]. Available: http://www.gutenberg.org/files/3207/3207-h/3207-h.htm. Accessed 22 July 2011
56. Friend, C.: Social Contract Theory. The Internet Encyclopedia of Philosophy (2004). http://www.iep.utm.edu/soc-cont/. Accessed 19 Oct 2009
57. Pateman, C.: The Sexual Contract. Stanford University Press, Palo Alto, CA (1988)
58. Baier, A.: Pilgrim's progress: review of David Gauthier, morals by agreement. Can. J. Philos. **18**(2), 315–330 (1988)
59. Baier, A.: Moral Prejudices. Harvard University Press, Cambridge, MA (1995)
60. Held, V.: Feminist Morality. University of Chicago Press, Chicago, IL (1993)
61. Held, V.: The Ethics of Care. Oxford University Press US, New York (2006)
62. Mills, C.W.: The Racial Contract. Cornell University Press (1997)
63. Debian: Debian Social Contract. Debian.org (2004). http://www.debian.org/social_contract. Accessed 26 Feb 2009
64. Knight, W.: Google's AI chief says forget Elon Musk's killer robots, and worry about bias in AI systems instead. MIT Technology Review, Oct 3, 2017. https://www.technologyreview.com/s/608986/forget-killer-robotsbias-is-the-real-ai-danger/. Accessed 8 Jan 2019
65. Ristroph, G.: Debian's democracy. In: Davies, T., Gangadharan, S.P. (eds.) Online Deliberation: Design, Research and Practice, pp. 207–212. Chicago, Illinois, Center for the Study of Language, USA (2009). [Online]. Available: http://www.press.uchicago.edu/presssite/metadata.epl?mode=synopsis&bookkey=5667101. Accessed 6 Sept 2011
66. Torvalds, L.: Torvalds. Harper-Collins, New York (2002)
67. Christie, N.: Conflicts as property. Br. J. Criminol. **17**(1), 1–15 (1977)
68. Electronic Frontier Foundation: Takedown Hall of Shame. Electronic Frontier Foundation (2009). http://www.eff.org/takedowns. Accessed 2 Nov 2009
69. Chambliss, W.J., Seidman, R.B.: Law, Order, and Power. Addison-Wesley Publishing Company, London (1982)
70. Weber, M.: Economy and Society: An Outline of Interpretive Sociology. University of California Press, Los Angeles (1978)
71. Marx, K.: Capital, vol. 1. Penguin Books Limited, London (1990)
72. Harris, M.: Glitch Capitalism: How Cheating AIs Explain Our Stagnant Present. Select All (2018). http://nymag.com/selectall/2018/04/malcolm-harris-on-glitch-capitalism-and-ai-logic.html. Accessed 16 May 2018
73. Sumner, C.: Reading Ideologies: An Investigation into the Marxist Theory of Ideology and Law. Academic Press, London (1979)
74. Locke, J.: Political Writings. Mentor, London (1993)
75. Moor, J.: The Turing Test: The Elusive Standard of Artificial Intelligence. Springer Science & Business Media, New York (2003)
76. Turing, A.M.: Computing machinery and intelligence. Mind **59**(236), 433–460 (1950)
77. Hoffman, M., Pfeifer, R.: The implications of embodiment for behavior and cognition: animal and robotic case studies. In: Tschacher, W., Bergomi, C. (eds.) The Implications of Embodiment: Cognition and Communication, Andrews UK Limited, Exeter (2015). [Online]. Available: https://arxiv.org/abs/1202.0440
78. Nilsson, N.J.: The Quest for Artificial Intelligence. Cambridge University Press, Cambridge (2009)

79. Brooks, R.: Cambrian Intelligence: The Early History of the New AI. 1st edn. A Bradford Book, Cambridge, Mass (1999)
80. Crevier, D.: AI: The Tumultuous History of the Search for Artificial Intelligence. Basic Books, New York (1993)
81. Newquist, H.P.: The Brain Makers, 1st edn. Ind: Sams, Indianapolis (1994)
82. Brooks, R.: A robust layered control system for a mobile robot. IEEE J. Robot. Autom. **2**(1), 14–23 (1986). https://doi.org/10.1109/JRA.1986.1087032
83. Pfeifer, R.: Embodied artificial intelligence. In: International Interdisciplinary Seminar on New Robotics, Evolution and Embodied Cognition, Lisbon, Nov 15, 2002. [Online]. Available: https://www.informatics.indiana.edu/rocha/publications/embrob/pfeifer.html. Accessed 10 Jan 2019
84. Renzenbrink, T.: Embodiment of Artificial Intelligence Improves Cognition. Elektor Ethics, Feb 9, 2012. https://www.elektormagazine.com/articles/embodiment-of-artificial-intelligence-improves-cognition. Accessed 10 Jan 2019
85. Zarkadakis, G.: Artificial intelligence and embodiment: does Alexa have a body? Medium, May 6, 2018. https://medium.com/@georgezarkadakis/artificial-intelligence-embodiment-does-alexa-have-a-body-d5b97521a201. Accessed 10 Jan 2019
86. Steels, L., Brooks, R.: The Artificial Life Route to Artificial Intelligence: Building Embodied, Situated Agents. Taylor & Francis, London, New York (1995)
87. Paradowski, M.B.: Developing embodied multisensory dialogue agents. In: AISB/IACAP 2012 Symposium, Birmingham, Nov 29, 2011. [Online]. Available: http://events.cs.bham.ac.uk/turing12/. Accessed 10 Jan 2019
88. Raz, J.: The Authority of Law: Essays on Law and Morality. OUP Oxford, Oxford (2009)
89. Foucault, M.: Discipline and Punish: The Birth of the Prison. Vintage Books, London (1995)
90. Chen, S.: AI research is in desperate need of an ethical watchdog. Wired (2017). [Online]. Available: https://www.wired.com/story/ai-research-is-in-desperate-need-of-an-ethical-watchdog/. Accessed 22 Sept 2018
91. Foucault, M: The History of Sexuality, vol. I. Penguin, repr, Harmondsworth, London (1981)
92. Foucault, M.: Disciplinary power and subjection. In: Lukes, S. (ed.) Power, NYU Press, New York (1986)
93. Foucault, M.: Power/Knowledge: Selected Interviews and Other Writings, 1972–1977. Pantheon Books, London (1980)
94. Foucault, M.: Power. Penguin, London (1980)
95. Foucault, M.: The subject and power. Crit. Inq. **8**(4), 777–795 (1982)
96. Galloway, A.R.: Protocol: How Control Exists After Decentralization. MIT Press, Cambridge, MA (2004)
97. Foucault, M.: The Birth of Biopolitics: Lectures at the Collège de France, 1978–1979. Pan Macmillan, London (2008)
98. Foucault, M.: Discipline and Punish: The Birth of the Prison. Pantheon, New York (1977)
99. Zimmer, M.: The panoptic gaze of web 2.0: how web 2.0 platforms act as infrastructure of dataveillance. Kulturpolitik (2) (2009). [Online]. Available: http://michaelzimmer.org/files/Zimmer%20Aalborg%20talk.pdf. Accessed 19 Jan 2010
100. European Commission: The anti-counterfeiting trade agreement (ACTA). Europa.EU, 2007. http://ec.europa.eu/trade/creating-opportunities/trade-topics/intellectual-property/anti-counterfeiting/. Accessed 30 Dec 2010
101. Lambert, J.: Statement on adoption of joint resolution on ACTA. Jean Lambert Green MEP for London, Nov 24, 2010. http://www.jeanlambertmep.org.uk/news_detail.php?id=620. Accessed 15 Dec 2010
102. Fuchs, C.: Social Networking Sites and the Surveillance Society. Verein zur Förderung der Integration der Informationswissenschaften, Vienna, Austria (2009)
103. Medosch, A.: Post-privacy or the politics of labour, intelligence and information. The Next Layer (2010). http://thenextlayer.org/node/1237. Accessed 19 Jan 2010
104. Wolf, C.: The Digital Millennium Copyright Act. Pike & Fischer—A BNA Company, Washington (2003)

105. Movius, L.B.: Surveillance, control, and privacy on the internet: challenges to democratic communication. J. Glob. Commun. **2**(1), 209–224 (2009)
106. Berkman Klein Center: Ethics and governance of AI (2018). https://cyber.harvard.edu/topics/ethics-and-governance-ai. Accessed 22 Sept 2018
107. Clark, J.: AI and ethics: people, robots and society. Washington Post (2018). http://www.washingtonpost.com/video/postlive/ai-and-ethics-people-robots-and-society/2018/03/20/ffdff6c2-2c5a-11e8-8dc9-3b51e028b845_video.html. Accessed 22 Sept 2018
108. Green, P.: Artificial intelligence and ethics, University Santa Clara (2017). https://www.scu.edu/ethics/all-about-ethics/artificial-intelligence-and-ethics/. Accessed 22 Sept 2018
109. Lufkin, B.: Why the biggest challenge facing AI is an ethical one. BBC (2017). http://www.bbc.com/future/story/20170307-the-ethical-challenge-facing-artificial-intelligence. Accessed 22 Sept 2018
110. Fitzpatrick, P.: The Mythology of Modern Law. Routledge, London (2002)
111. West, R.: Narrative, Authority, and Law. University of Michigan Press, Michigan, MI (1993)
112. Adams, J.N., Brownsword, R.: Understanding Law. Sweet & Maxwell, New York (2006)
113. Courtland, R.: Bias detectives: the researchers striving to make algorithms fair. Nature (2018). http://www.nature.com/articles/d41586-018-05469-3. Accessed 6 July 2018
114. Fry, H.: We hold people with power to account. Why not algorithms? The Guardian (2018). [Online]. Available: https://web.archive.org/web/20190102194739/https://www.theguardian.com/commentisfree/2018/sep/17/power-algorithms-technology-regulate. Accessed 17 Sept 2018
115. Ito, J.: Resisting reduction: a manifesto. J. Des. Sci. (2017). https://doi.org/10.21428/8f7503e4
116. Bridle, J.: Rise of the machines: has technology evolved beyond our control? The Guardian (2018). [Online]. Available: https://web.archive.org/web/20190111222310/https://www.theguardian.com/books/2018/jun/15/rise-of-the-machines-has-technology-evolved-beyond-our-control. Accessed 15 June 2018
117. Singh, P.J.: AI superpower or client nation? The Hindu (2018). [Online]. Available: https://www.thehindu.com/opinion/op-ed/ai-superpower-or-client-nation/article24523017.ece. Accessed 14 Aug 2018
118. Keeble, R.: Ethics for Journalists. Routledge, London (2008)
119. Kieran, M.: Media Ethics. Psychology Press, London (1998)
120. Etzioni, O.: How to regulate artificial intelligence. The New York Times, Jan 20, 2018. [Online]. Available: https://www.nytimes.com/2017/09/01/opinion/artificial-intelligence-regulations-rules.html. Accessed 22 Sept 2018
121. Goel, A.: Ethics and artificial intelligence. The New York Times, Dec 22, 2017. [Online]. Available: https://www.nytimes.com/2017/09/14/opinion/artificial-intelligence.html. Accessed 22 Sept 2018
122. Vidmar, N., Miller, D.T.: Socialpsychological processes underlying attitudes toward legal punishment. Law Soc. Rev. 565–602 (1980)
123. Wenzel, M., Okimoto, T.G.: How acts of forgiveness restore a sense of justice: addressing status/power and value concerns raised by transgressions. Eur. J. Soc. Psychol. **40**(3), 401–417 (2010)
124. Bostrom, N., Yudkowsky, E.: The ethics of artificial intelligence. Camb. Handb. Artif. Intell. **316**, 334 (2014)
125. Frankish, K., Ramsey, W.M.: The Cambridge Handbook of Artificial Intelligence. Cambridge University Press, Cambridge (2014)
126. Fionda, J.: Devils and Angels: Youth Policy and Crime. Hart, London (2005)
127. de Tocqueville, A.: Democracy in America. Library of America, Washington, DC (2004)
128. Montti, R.: Google's "don't be evil" no longer prefaces code of conduct. Search Engine J. (2018). https://www.searchenginejournal.com/google-dont-be-evil/254019/. Accessed 22 Sept 2018
129. Duff, R.A.: Punishment, Communication, and Community. Oxford University Press, Oxford (2003)
130. Durkheim, E.: The Rules of Sociological Method. Vani Prakashan, New Delhi (1960)

131. Braithwaite, J.: Restorative justice and a better future. In: McLaughlin, E., Hughes, G. (eds.) Restorative Justice: Critical Issues, pp. 54–67. SAGE, London (2003)
132. Braithwaite, J.: Crime, Shame and Reintegration. Cambridge University Press, Cambridge (1989)
133. Sample, I.: Technologist Vivienne Ming: "AI is a human right". The Guardian, Dec 7, 2018. [Online]. Available: https://web.archive.org/web/20190111125507/https://www.theguardian.com/technology/2018/dec/07/technologist-vivienne-ming-ai-inequality-silicon-valley. Accessed 09 Dec 2018
134. Conn, A.: Podcast: law and ethics of artificial intelligence. Future of Life Institute (2017). https://futureoflife.org/2017/03/31/podcast-law-ethics-artificial-intelligence/. Accessed 22 Sept 2018
135. Rawnsley, A.: Madeleine Albright: "the things that are happening are genuinely, seriously bad". The Guardian, July 8, 2018. [Online]. Available: https://web.archive.org/web/20190106193657/https://www.theguardian.com/books/2018/jul/08/madeleine-albright-fascism-is-not-an-ideology-its-a-method-interview-fascism-a-warning. Accessed 9 July 2018
136. Haraway, D.: A Cyborg Manifesto. Social. Rev. **15**(2) (1985). [Online]. Available: http://www.stanford.edu/dept/HPS/Haraway/CyborgManifesto.html. Accessed 14 Nov 2009
137. Thompson, C.: The cyborg advantage. Wired **18**(4) (2010). [Online]. Available: https://www.wired.com/2010/03/st-thompson-cyborgs/. Accessed 22 Sept 2018
138. Hipp, J., et al.: Computer aided diagnostic tools aim to empower rather than replace pathologists: lessons learned from computational chess. J. Pathol. Inf. **2** (2011). https://doi.org/10.4103/2153-3539.82050
139. Baggini, J.: Memo to those seeking to live for ever: eternal life would be deathly dull | Julian Baggini. The Guardian, July 8, 2018. [Online]. Available: https://web.archive.org/web/20181225111455/https://www.theguardian.com/commentisfree/2018/jul/08/live-for-ever-eternal-life-deathly-dull-immortality. Accessed 9 July 2018

Training Social Skills in Virtual Reality Machine Learning as a Process of Co-Creation

Klaus Neundlinger, Michael Mühlegger, Simone Kriglstein, Thomas Layer-Wagner, and Georg Regal

Abstract The training of social skills in interactive Virtual Reality scenes has become an issue of rising interest in research and development. This is also the case of the project Virtual Skills Lab that deals with the question how the potential of such a training can be applied to the context of organizations. In a virtual office scene, the interaction between the player and the computer-generated agent is based on speech recognition and conversational Artificial Intelligence (AI). While studies and experiments have shown that humans are able to treat agents in Virtual Reality as if they were real persons, it is still impossible to establish truly spontaneous forms of interaction by combining speech recognition, conversational AI and Virtual Reality. Hence, existing applications for social skills training in Virtual Reality focus on the possibility to literally take over the other's perspective or the ability to overcome social anxieties. Yet, social interaction in the workplace is always embedded in a given organizational culture and therefore characterized by behavioral patterns. In this sense, a Virtual Reality training tool for social interaction in the workplace can be enriched by design elements that enable machine learning in terms of pattern recognition. The paper outlines how the project team has dealt with the existing limitations of conversational AI in the process of working out the storyline and what design elements can be introduced in the Virtual Reality interaction in order to enable the machinic recognition of patterns of organizational culture.

Keywords Virtual reality · Social skills training · Co-creative research

---

K. Neundlinger (✉)
Institute for Cultural Excellence, Oberalm, Austria
e-mail: klaus.neundlinger@institute-ce.at

M. Mühlegger
Karl Landsteiner Privatuniversität für Gesundheitswissenschaften, Krems, Austria

S. Kriglstein · G. Regal
Austrian Institute of Technology, Vienna, Austria

T. Layer-Wagner
Polycular, Hallein, Austria

A. Dingli et al. (eds.), *Disruptive Technologies in Media, Arts and Design*, Lecture Notes in Networks and Systems 382, https://doi.org/10.1007/978-3-030-93780-5_10

1 The Virtual Skills Lab Project

Immersive Virtual Reality (VR) training has been proposed for a wide range of usage contexts. It roots in combat or flight simulators, one of the earliest immersive training applications, and has been applied to various other domains like professional skills training, training for first responders, workplace safety trainings, etc. In the last years, focus has been also laid on rather complex training issues dealing with emotions, social interaction and communication, like empathy towards others, overcoming the anxiety to speak in public, dealing with diversity, etc. Thus, a question that increasingly occupies researchers and developers is how interactive VR experiences can be compared to social interaction between humans, in order to fulfill the requirements of an effective training based on the simulation of real-world interaction. Will it ever be possible to design a machine intelligence that dialogues with humans, to act and react not only regarding to the information processed and conveyed in the course of a conversation, but also regarding the social, cultural and emotional layers of communication?

In order to achieve a certain degree of spontaneity between humans and virtual agents, human–computer interaction has to be supported by various forms of machine learning, such as chatbots, speech or emotion recognition. In this paper we present a particular perspective on machine learning in the context of interactive VR that is based on findings of the Virtual Skills Lab (VSL) project, an interdisciplinary research and development project. The project aims at developing a research prototype for the training of social skills using Virtual Reality. For this purpose, a scene is being developed in which a player, using a head-mounted display, finds herself in a virtual office space. In the scene, the user takes the role of a manager interacting with an employee in a stressful situation.

1.1 The Ideas Lab

The project idea was developed during a 5 days' "Ideas Lab" in September 2018 in which representatives from research institutions and firms developed possible project ideas around the question of how human–machine interaction can and should be designed in the future. One of the principal goals of the "Ideas Lab" was to facilitate interdisciplinarity taken seriously throughout the whole research and development process. For this reason, the project unites scholars engaged in basic and applied research as well as practitioners and developers from the areas of corporate culture and learning technology. Hence, not only the objectives and methods are relevant for the project's success, but in particular also the process, for its deliberate openness.

As mentioned above, the task of the "Ideas Lab" was to develop project ideas emerging from the question how humans and machines can and should possibly interact in the future. Within the creative process, the idea was brought up to use data collected in more than 100 workshops, carried out by the institute for cultural excellence (ice). The aim of these workshops was to analyze the quality of daily social interaction within teams and organizations and to develop it towards an appreciative work culture.

The data set consisted fundamentally in assessments made by employees and managers concerning the quality of working together in their teams, in projects or departments. In these assessments, workshop participants expressed their experiences and expectations on how people should or should not interact with each other in projects and teams. One of the tasks of our research project is to relate these data to the possibilities of machine learning and to transfer them into Virtual Reality learning experiences by storytelling. Will it be possible to use these data with the aim to initiate processes of pattern recognition, as they are typical for machine learning, in the realm of human social behavior?

1.2 Training Social Interaction to Improve Organizational Culture

Corporate culture, on the one hand, can be conceived of as a more or less common, more or less conscious set of basic assumptions, ideas, values and behavior patterns shared by the people working in an organization [1]. Corporate culture, in this sense, provides the members of an organization with orientation and stability for the projects and processes that are carried out in everyday cooperation [2]. On the other hand, corporate culture is nothing fixed for all times. It is reproduced, but also continuously re-shaped and modified by the single employees and their reciprocal behavior. In other words, employees and managers at any level of a company constantly contribute to the working atmosphere and can enhance (or deteriorate) it by developing (or not) their social skills, i.e., their ability to interact with each other in a respectful way [3, 4]. What we had as a starting point was a data set consisting of experience-based assessments on the quality of working together in a series of companies, as well as statements concerning the expectations, expressed by employees and managers, on what "good collaboration", i.e., socially skilled interaction in the workplace, should be like.

Against this background, the original idea of the Virtual Skills Lab was formulated. Our goal was to use Virtual Reality as an instrument offering the possibility to people working in organizations (especially to the midlevel management) to train their individual social skills by interacting with an agent representing an employee in an office environment, i.e., by means of experiential learning.

In our research, we aim at answering the question what could be the advantages of transferring the interactive training of social skills to VR, compared to role play and other forms of simulation. There are some hypotheses we aim to evaluate by constructing our prototype. VR is deemed to offer additional opportunities to the learner, e.g.,

- to train social interaction in a safe environment (without harming other persons or being harmed),
- to be able to infinitely repeat the interaction,
- to experience perspective taking instead of only imagining it,
- to design and measure different degrees of intensity and difficulty of the interaction,
- to modulate the agent and the environment in terms of the degree of abstractness of the representation.

2 VR Training of Social Skills

Social interaction between humans is a highly complex process, determined not only by verbal messages and their interpretation, but also by many other dimensions such as nonverbal communication via facial expression, gestures and so on. It is characterized by ambiguity and by unconscious mechanisms that have been studied by social and group psychology for a long time. In addition, interaction takes place in a highly structured social environment. As human beings, we learn to behave according to specific cultural codes, explicit or implicit hierarchies and other factors that influence the way we relate to each other in a very subtle way [5, 6].

2.1 Social Interaction in the Workplace

The complexity of social interaction and communication between humans has been gaining importance also in the context of work and organizations in the last decades [4, 7]. This is due to the fact that the structuring of cooperation in firms has shifted from the hierarchical mode of order-and-execution, typical for industrial work, to a greater variety of interaction types, more typical for knowledge intensive and service-based forms of value production [8, 9]. Within teams and projects, processes of deliberating, reflecting, developing ideas, plans and tasks at eye level occur at a regular basis. Having to interact more closely with others in terms of exchanging information, knowledge or perspectives has led to a higher personal and emotional involvement in organizational processes. People experience themselves and their colleagues not only as fulfilling tasks assigned them by their supervisors, but also as shaping and constructing cooperation by their way of communicating [10–12]. Nowadays, effective teamwork depends on the social skills of the persons involved in the same measure in that it depends on their professional skills and knowledge.

As a consequence, there has been an increasing necessity and demand regarding the reflection on, and the training of, social skills in organisational contexts. Socially skilled behavior is studied and measured by sociology [3] and organizational psychology. Psychologists have set up systematic approaches in order to diagnose these person-related skills for job training and selection purposes [13]. Besides methods like self and external assessment or 360° feedback, also interactive forms such as role plays and simulations have been used for social skills training. Once social skills are object to training measures and methods, the complexity of spontaneous social interaction is reduced to, and focused on, determined aspects. This applies also to the more interactive forms although they are considered to be "closer" to practice, since they contain experiential elements, in opposition to merely reflexive or discursive forms of learning.

As Goffman [6] put it in his famous analysis on everyday social interaction, we all play theater. At any time. No interaction can take place without being staged, without a symbolic framework. Roles and functions are symbolic reference systems that help us get through our daily life. As human beings, we learn to behave according to the expectations and conditions given not only by the specific personal traits of our interlocutors, but also by their—and our own—role and function within a given situation or institutional setting. Social reality is always mediated by symbolic structures and forms. The ways of presenting and re-presenting oneself in private and public life have been multiplied by communication technologies and the internet. Role plays, thus, are to be considered a reduced form of how we behave and interact at any given time and in any given situation in everyday life. By reducing the interaction and highlighting a specific didactic aspect, role plays can be made fruitful as learning experiences.

2.2 VR as Environment for Training Social Interaction

The aspect of experiencing something as real or realistic occurs not only in role plays, but also in Virtual Reality (VR), with its promise to the users of "being there", of being immersed and present in a virtual environment and experiencing oneself as actor [14]. In several empirical studies (based on prototypes), the appropriateness and effectiveness of training social skills in VR has been highlighted [15–18]. In the meantime, also a certain number of artistic, therapeutic and commercial applications has been developed. In diverse settings, people have the possibility to overcome their anxiety to speak in public, they learn to empathically react to employees with difficult characters or to social welfare recipients represented by agents, having the possibility of experiencing the own way of interacting from the point of view of the virtual agent afterwards. Thus, while on the one hand VR training settings are designed to enable the player to cope with internal insecurities and to interact in a self-confident way, on the other hand VR provides the possibility to walk in another's shoes, i.e., to take the other's perspective in a stricter sense of the term [19]. Both aspects are crucial for the development and training of social skills [13].

Yet, there is still a long way to go until more spontaneous ways of interacting will be possible with agents based on speech recognition and conversational artificial intelligence (AI). Looking at the existing research and VR applications involving social interaction one can state that those are appropriate technologies in the sense that users tend to treat agents like human beings. They get cognitively and emotionally involved in the interaction, albeit still based to a great extent on their imagination and not so much on intriguing, surprising or profound types of conversation [15].

In the field of conversational AI, progress is being undoubtedly achieved in terms of consistency, specificity and empathy [20–22]. It depends on the development of speech recognition, the major interface between human and machine in terms of spontaneous natural language-based interaction, if this progress can sooner or later be fruitfully applied in VR scenes. Another important element in this sense is emotion recognition, a research field in which progress is being made, but it is still difficult for machine learning to distinguish more subtle forms of emotional expression, such as vocal irony [23].

2.3 Solutions for VR Soft Skills Training and Perspective Taking

There are already some applications for the training of social skills commercially available, for instance the VR training applications for leadership, diversity and inclusion, customer service or sales developed by the company Mursion [24]. Other companies, like Talespin, Strivr or Equal Reality, offer trainings for managers and employees dealing with socially and emotionally demanding situations [25–27]. In some cases, such as in the VR trainings of Bodyswaps [28], eye tracking and other biofeedback devices are used to measure not only the verbal, but also the nonverbal reactions of the player to the scene and to the virtual agent. After having run through an interactive scene, the players are given feedback regarding their emotional engagement displayed not only by the use of words, but also by eye contact and other forms of nonverbal, bodily expression.

Furthermore, several projects in academic research and arts investigate into the possibility that VR offers to enable human beings to better understand other people's feelings and views, like the scientific experiments of the Virtual Human Interaction Lab on empathy and perspective taking [29], the artistic experiments of the BeAnotherLab [30], enabling people to go through the bodily experience of being another person, or the project Manic VR by the artist Kalina Bertin that invites people to experience the mental and emotional states of persons suffering from bipolar disorder [31].

All these applications aim at training the capacity of humans to interact in an empathic way towards others. But can the machine itself take over a more active role that is perceived by their human interaction partners as adequate and empathic? There are approaches that, in addition to speech recognition and eye tracking, use the machine-based recognition of facial expressions and movements, aiming at enabling the machine to detect the players' emotional state [32]. Still, this type of human–computer interaction is not possible in VR settings based on the use of head-mounted displays, i.e., the player interacts with a 2-dimensional agent appearing on a computer screen. In other words, the most important aspects of VR in terms of user experience, immersion and presence, are less effective in this setting.

3 The Co-Creation of a Social Skills Training

Our research project is based not only on interdisciplinarity, but also on a participatory approach. That means that we co-create our prototype with potential users and iterate this participatory process by integrating more feedback from focus groups during the whole project cycle. For this reason, we held a series of workshops with a group of midlevel managers working in an international corporation that was founded in Austria. In the workshops, we asked the managers to reflect on difficult situations in daily interaction with their collaborators as well as on their ideas and convictions about "good collaboration".

Amongst the topics that emerged from these workshops, the problem of "saying no in a respectful (appreciative) way" was chosen as exemplary situation to be developed as an interactive scene. On the basis of the experiences reported by the managers, a screenwriter worked out the scene which was further developed and modified together with the managers.

By carrying out the workshops, we continued to collect qualitative data concerning the question the institute for cultural excellence had been dealing with in the past, i.e., what good collaboration in the sense of an appreciative, respectful work culture should be like. As already stated, work culture is the result of collective interaction patterns which are continuously reproduced or reshaped, i.e., transformed. Being part of teams and organizations, people learn to recognize these patterns and to behave accordingly. Yet, they are also able to critically reflect on these patterns and change them by acting differently or opposing to dysfunctional ways of interacting. The data that form the basis of our research are assessments about how people perceive and value the collaboration quality. That is, these assessments reflect their more or less conscious ways of dealing with behavioral patterns.

With respect to machine learning, the challenge of any prototyping in this area is to find a way to enable the computer to recognize such patterns, i.e., to learn something about how people think about and conceive of "good collaboration". The challenge of human–computer interaction in the case of training social interaction in the workplace is, so to say, to extend the co-creation process to the algorithm. While human learners experience the office scene as a way to further develop their ability to reflect on and transform their social behavior, machine learning regards the recognition of linguistic patterns valued as "respectful" or "appreciative" in the context of working together in organizations. Similarly, also the possibility of a machine-based recognition of patterns valued as "disrespectful", "discriminating" or "derogatory" could be used to train "negative" scenarios in which a conflict escalates, maybe followed by instructions or tutorials how one should not interact and communicate in certain situations.

Summing up, for the development of the VR scene we used two sources. One was a set of routine data retrieved regularly in workshops with teams that aimed at analyzing and improving the quality of working together. People were asked to describe their daily interaction and the workplace atmosphere by referring to a set of given categories. In addition, they were asked to express what "good collaboration" meant for them. The other source was a series of workshops with midlevel managers in which we co-created the exemplary workplace scene the topic of which was "how to decline an employee's request in an appreciative way".

3.1 Dramaturgic and Technical Challenges

The challenge deriving from these workshops was to translate this topic into an interesting interactive experience. Players should experience the tension between trying to be supportive towards their employees and having to reject a legitimate request. They should not only cognitively, but also emotionally, face the contradiction between the idealistic self-image of a supportive manager and the actual lacking of time resources in order to behave accordingly at any given moment, especially in stressful situations in which it is difficult to set the right priorities. There are many situations in everyday office life apt to exemplify this underlying tension, even small conversations between two meetings or in the office corridor. Yet, as simple as a dialogue of one person making a request and the other rejecting it may seem, it involves the abovementioned layers of sociality and emotionality that human beings almost naturally perceive, but that are not so easily to be detected by a virtual interlocutor. The challenge, thus, was twofold: on the one hand, from a dramaturgical standpoint we had to construct a dialogue that should be representative for many different working contexts, but not too generic to evoke emotional engagement. On the other hand, regarding the VR experience, we had to deal with a series of limitations that still make it difficult for a speech and emotion recognition system to engage in an authentic conversation. In other words, we had to find a way to make machine learning part of the co-creation process by combining the learning experience of the users with that of the machine.

Our solution, which we will outline in the next two sections, consisted in keeping the dialogue linear without leaving any possibility to the player to resolve the conflict with the virtual agent positively at least in a first round. In a second round, players are given the possibility to explore the various virtual layers that constitute any linear conversation and to train their reflexive capabilities regarding social interaction. On the other hand, we decided to insert interactive elements that enable machine learning so that it is possible to build up neural networks for human–machine interaction in the realm of training social skills.

In Sect. 4, we will briefly describe the office scene dealing with "saying no in an appreciative way" that we have developed (as first version of the prototype) and then discuss the various aspects and problems of constructing it in VR and combining it with conversational AI. In Sect. 5, we will present our solution for the VR office scene in more detail and make some design proposals for how to combine the human learning experience with machine learning in a co-creative process.

4 The Office Scene

The scene is organized as a role play, with the user's avatar appearing in VR as a midlevel manager who has a team of about ten collaborators under his (or her) supervision. One of the collaborators named Mira Horvath enters the manager's office and reminds the manager of their scheduled meeting, starting right in five minutes (Fig. 1). The player learns that he (or she) is in a conflicting situation, as a text message arrives, asking him (her) to attend another unplanned meeting of higher priority at the same time. The goal of the training is to find an appreciative way of refusing to start the meeting with Mira. In the simulation, Mira won't accept the refusal and after a short discussion she leaves the manager's office in anger.

4.1 How Real Does Virtual Need to Be?

Due to the rapid technological evolution in the last years, VR provides a realistic simulation environment, apt not only for games, but also for training situations. Virtual human agents can be designed and programmed in many different ways and can move and look quite realistic. As we outlined above, human social interaction is based on verbal and nonverbal cues. This means that, in order to be accepted as an

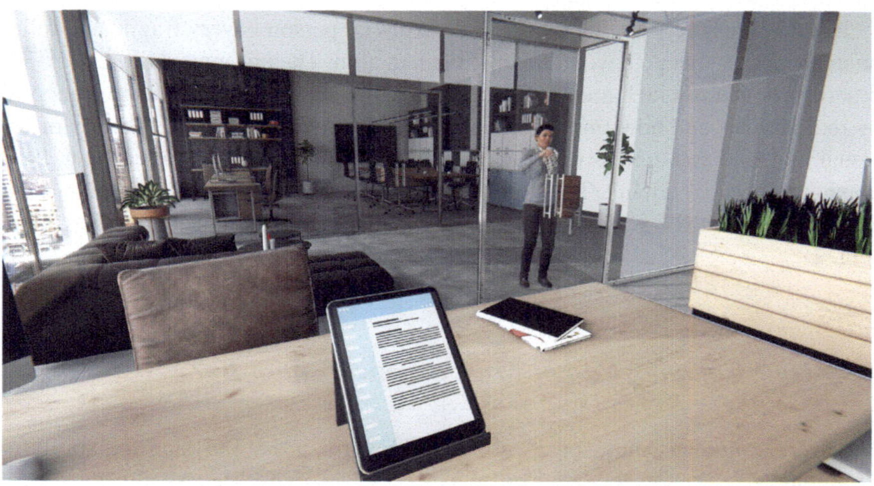

Fig. 1 Screen shot from the Virtual Skills Lab prototype. Mira Horvath enters the office

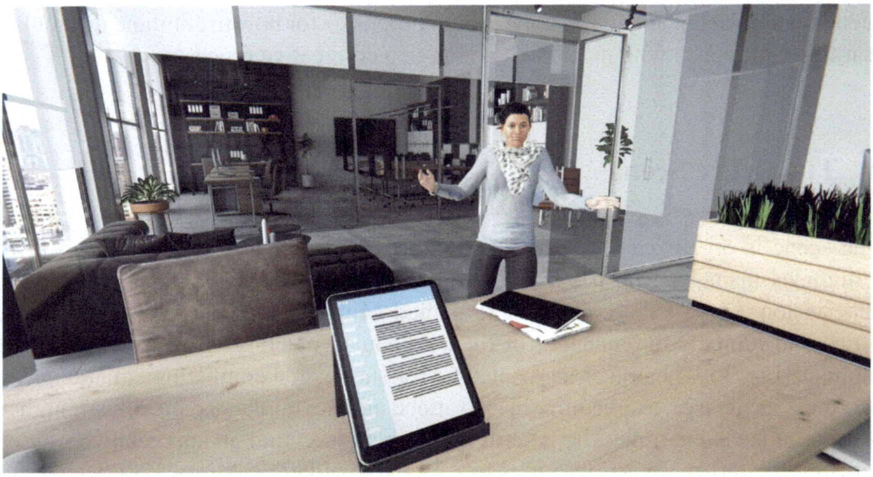

Fig. 2 Screen shot from the Virtual Skills Lab prototype. Mira explains why she urgently needs a decision from her superior

interaction partner by a human user, the artificial human has to talk and understand not only human language, but perceive and actively use also gestures and facial expressions. Voice assistants on smartphones like Apple's Siri or Amazon's Alexa are used every day by millions of users and establish a more natural way of human–computer interaction. As VR tries to simulate reality as detailed as possible, the user input needs to be intuitive, possibly without the use of any additional input devices like a keyboard, computer mouse or other controllers. Therefore, voice recognition and natural language processing had to be integrated into the training for the user interaction.

But this illusion of interacting with an artificial person is very fragile and relies on many details that go beyond linguistic interaction. The uncanny valley effect [33] describes how artificial humans, especially at a high degree of realism, are perceived by real humans and how tiny detail deviations destroy the illusion. In terms of the visual look and movements of the virtual human, we decided not to reach out for a maximum level of fidelity. That means that Mira looks realistic, but it is also clear to the player that she is no real human, in order to avoid the uncanny valley effect (Fig. 2).

4.2 *How Complex Can or Must a Conversation Be?*

The average human being speaks about 7000 words a day [34] and the ability to communicate as a species is a central aspect of our daily life. As already stated, the exchange of information between human beings comprises verbal and nonverbal communication components. A regular conversation has an overall topic and each conversation participant has an intent. As any conversation is a highly dynamic process, this intent and context can change and be adapted many times during the conversation. The verbal exchange can be viewed as an act of synchronization between multiple entities at which information has to be simultaneously received and produced. Humans learn the ability to process verbal information early on. The process of language acquisition starts at a prenatal stage and keeps on adapting and changing during the whole life span [35].

According to the four-sides-of-communication model [36], verbal information splits into the four separate layers: factual information, self-revelation of the sender, appeal towards the receiver and the relationship between the two conversational partners. Hence, receiving a message is a multi-layered act of "reading between the lines". That means that the interpretation of spoken words is essentially ambiguous and depends on the chosen communication layer. The decision which layer of communication to focus on depends on the context of the conversation and the intrapersonal emotional state of the sender and the receiver. Multiple other factors like diplomatic strategies, cultural conventions, intonation and timing, humor or sarcasm, can also influence the way information might be interpreted.

As verbal information has a large scope of interpretation, other sources of information can or must be integrated into the process of interpretation such as nonverbal expression [37]. The facial expression of the conversational partner, posture or gestures can help the recipient to select the intended meaning of a message (Fig. 3). But similar to verbal information, the nonverbal counterpart is ambiguous and culturally specific as well. Just to give some well-known examples: in Western cultures, the shaking of the head can be interpreted as a "No" or a form of disagreement. In the Indian culture, the shaking of the head has the opposite meaning and symbolizes consent.

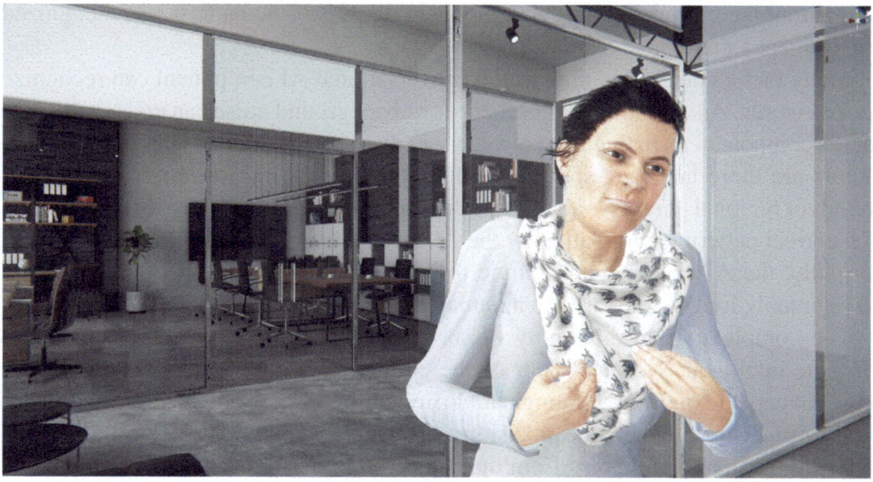

Fig. 3 Screen shot from the Virtual Skills Lab prototype. Mira's facial expression and gesture while she tells her superior why she is annoyed about the postponement of the meeting

According to the respective recipient's experience, the expected intent, the given cultural context or other situational factors, all this information has to be processed and integrated to form meaning during a conversation between humans. The output of this complex process builds the foundation of a recipient's response. Verbal and nonverbal signs and signals need to be decrypted and interpreted under many aspects. In addition, conversations build up quickly and can refer to prior conversations forming even deeper levels of complexity. It is not possible to embed a single VR interaction into a complex network of references, but it could be a future challenge to design interactive experiences for the training of social interaction in the form of stories, i.e., to construct more sophisticated narratives in which characters and stories, like in role plays, evolve, so that also the virtual agent must be able to refer to prior conversations and situations.

4.3 How Mira Deals with Social and Conversational Complexity

Mira, the artificial conversational agent, or more precisely the AI that generates Mira's reactions, receives words pronounced by the players as sound waves. The speech-to-text service transforms these sound waves into machine-readable information, which can be analyzed by a trained neural network of the conversational AI component specialized to identify patterns. At this stage, Mira still can't make decisions based on the given input. To do so, the AI component must be able to recognize which elements of the spoken words are relevant and if the desired keywords appear in their correct order. Specified keywords or patterns get analyzed by a neural network which calculates their likelihood in terms of the intended word-meaning. Based on this estimation, the conversational AI component decides if the pattern was recognized and triggers a predefined response.

Many factors determine how well a conversational AI component can recognize spoken words. A low audio volume, echoes or background noise can make it difficult for the component to distinguish between specified sound patterns. Other challenging factors are more language specific. Different accents, a mix of languages, similar sounds or regional phrases—all of these disruptive factors can decrease the likelihood of a correct pattern identification by the AI. Even if every word could be recognized correctly, the AI response needs time to process all of the given information and takes notably longer compared to a human being.

The complexity and ambiguity of social interaction is not an easy task for a machine intelligence to deal with. At the current state of technological evolution, AI is not capable of differentiating paradox emotions or of easily identifying humor, irony or sarcasm with the same precision as a human counterpart [23]. During conversations, humans often digress from the subject in order to provide further explanations, returning only after a while to it. Such a thematic excurse can even result in a sub-dialogue that has no apparent relation to the original conversation. In order to be recognized as an equal conversation partner, conversational AI must be capable of keeping track of these changes and needs to guide the conversation branch back to the root topic without appearing rude.

If we take a closer look, the abovementioned challenges are not specific to AI and appear to be challenging for real humans as well. Both, AI and humans, have to face ambiguity in meaning which frequently leads to misunderstanding. In other words, also humans learn and develop conversational strategies for dealing with utterances they do not understand. Such strategies can be overt, like asking: "Pardon, what did you say?" or "Sorry, I didn't get the point, could you explain this to me?". But humans can also try to conceal their lack of understanding by continuing the conversation and trying to find out what the interlocutor meant indirectly, deducing it from the context. They can also choose a middle way by asking questions that do not directly reveal the fact that they are unsure about the other speaker's intent, disguising their uncertainty as interest in "learning more about this topic".

There are in fact many nuances, many ways of deciphering the other's intended meaning, as well as there are many subtle ways of constructing meaning together and sharing it throughout the interaction. This continuous creating of meaning as social action could be seen as the fundamental difference between human–human conversation and human–machine conversation. Yet, we have stated that interaction, especially when it is framed by a social context such as an organization, evolves in form of patterns. When communication happens, it creates always a specific, unforeseeable dynamic, but at the same time it is also scripted. It is like a text or a score that follows rules, patterns and schemes [4]. A conversation between a supervisor and their employee, between a parent and their child, between a teacher and their students, for example, evolves against the background of a specific role and power distribution. In other words, it evolves not only because of an interaction dynamic, but also because there is already a basic understanding of the roles, which are represented and enacted by the humans that embody them in the given situation. Still, regarding the way humans understand and embody the role they represent in a given situation there are many variants. Humans are able to perceive and reflect these possible variations regarding the understanding of one's (and the other's) role, a capability they develop during their socialization on the basis of their experiences, especially in the professional field. To vary and change behavior and behavioral patterns is demanding for them, but they have the possibility to do so. It is part of their individual history [38], an essential trait that distinguishes them from machines.

Just as the roles, also the topic follows a determined script in its evolution during the conversation. It is not by chance that the English noun "topic" derives from the Greek word *topos*, which means "place", or "commonplace". Ancient Greek philosophers already systematized the common themes, concepts and arguments that form a discourse or a discussion in the oratory disciplines of rhetoric and dialectics, denominating them *topoi*. And "subject", which is another synonymous expression for "theme", derives from the Latin word *subjectum*, which means "underlying". When we talk about any given theme, we move between given commonplaces in a discursive territory that does not only emerge by the words we say and the thoughts we express, but also shapes and channels our utterances and the course of the conversation from the beginning. The variants emerging from conversational dynamics can be analyzed and are accessible to machine recognition, as the computer-linguistic branches of text and argument mining show [39]. Furthermore, as mentioned above, advances in conversational AI are being made regarding not only the rational structure of argumentation and information, but also with respect to linguistic patterns perceived as empathic or meaningful reactions by humans.

In the course of their life, humans develop the ability to adapt their communicative strategies during the conversation and use their verbal flexibility to overcome challenging situations. Humans do so by reflecting on the situations they happen to experience and by building on these reflections. The ability to ask the right questions in a given situation allows us human beings to seize the indented meaning, i.e., to participate in the co-creation of a not only functional, but meaningful dialogue. Thus, although for Mira, in order to participate in a co-creative interaction with a human, it is not necessary to understand every single detail of the human's utterances, she has to develop communicative strategies that enable her to act according to a specific socially, culturally and interpersonally determined script. This would be the social skill she has to acquire. Translated into a technological problem, the question is how AI, by means of pattern recognition, can develop the ability of dealing with social uncertainty and ambiguity [40].

5 How We Are Dealing with Mira's Limitations

Returning to the Virtual Skills Lab, initially the conversation with Mira was designed as a complex tree structure, i.e., like most chatbots are structured. Depending on the given answer of the trainee, the conversational AI component should detect the context-specific meaning within the spoken words to decide which scene branch has to be triggered next. Still, conversational AI cannot yet fully deliver the flexibility and the dynamic aspects of a conversation between real human beings, and any attempt to emulate a natural conversation would not have led to a satisfying result in terms of user experience. Therefore, we decided to go the opposite way, by limiting the degree of freedom during the artificial conversation with Mira to a minimum. Instead of trying to get close to a free, unguided dialogue by constructing a fine ramified tree structure, we decided to pre-determine the options of how to answer to Mira's statements, and to limit them drastically for the first prototype. The trainee has to choose between two or three predefined answers and reads out the chosen option. The spoken answer is processed and the conversational AI detects specific keywords in form of a pattern recognition process to trigger the next step of the dialogue. Due to the linear design of the scenario, it is not important for the evolution of the dialogue which answer was chosen. In other words, the design induces the illusion of "freedom of choice", but limits the variability of the possible answers, leading in any case to the same outcome: Mira is visibly disappointed and leaves the office in anger (Fig. 4).

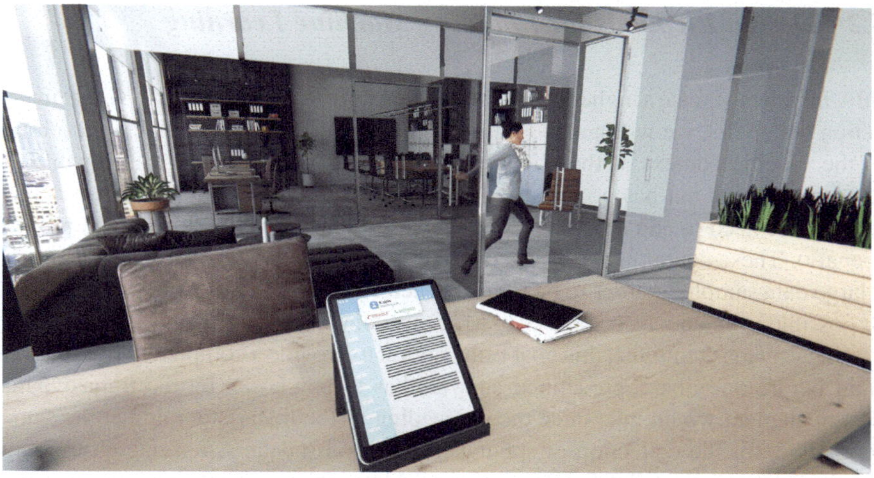

Fig. 4 Screen shot from the Virtual Skills Lab prototype. Mira leaves the office in anger

5.1 Reflecting on the Virtual Character of Social Interaction

Still, the player is given a second chance, so to say. While going through the dialogue once more, the player is provided with additional information regarding Mira and other circumstances of the scene. In this way, the user is prompted to reflect on the various levels and layers that determine and shape social interaction. In other words, despite the limited actual dynamic he or she has experienced in the scene, the trainee is enabled to reflect on the different factors and aspects that virtually influence social interaction, like limited information, taken for granted beliefs and attitudes, the lacking capability of taking the other's perspective and so on. Instead of experiencing a social interaction that is based on empathy in an emotional sense, the trainee is prompted to reflect on the structure of socially skilled, empathic behavior, experiencing the different aspects and layers of social interaction rather than its linear dynamic. In other words, the player does not run through a training in which he (she) learns how to "mechanically" react in a given, typical situation. Instead, the learning process targets the learner's reflexive competence to contemplate and evaluate social interaction, to evolve a higher level of awareness and to improve the capability to react and interact even in novel situations.

An additional benefit for the prototyping process is the reduction of development costs, compared to the task of building a complex tree structure with multiple optional scenario branches which would require a huge set of predefined reactions for Mira. The downside of this approach is its small degree of freedom and spontaneity that is only compensated by the possibility to run through the scene once more on a meta-level.

5.2 Design Elements for Enabling Machine Learning

Still, in order to create a higher degree of freedom during the first round of verbal interaction with Mira, we will need to train the conversational AI component according to the specific context of the VR scene. The text-to-speech service is capable of detecting words, but has no information about our scenario. To provide this information, we need to add other layers of specialized neural networks tailored for our scenario. These neural networks need to detect context-specific actions and require a large data sample in order to be trained. Therefore, we need to build up a database of possible spontaneous responses to Mira's statements.

One way to accomplish this goal is the option to provide a completely free answer right before one of the predefined answers has to be chosen by the trainee. This design element would guarantee the proceeding of the linear story and at the same time allow for the collection of spontaneous response data. After enough training data have been collected, this information could be processed by supervised learning algorithms and create an artificial neural network capable of detecting more context-specific patterns. With such an adapted AI, Mira could handle a wider range of responses and deal with more spontaneous answers. However, it is still possible that an unforeseeable answer occurs which confuses the AI. In such a case, Mira needs backup strategies to handle similar to a human that does not understand an utterance in an ongoing conversation. A repertoire of phrases like "I don't get it, could you explain the last part again please?" could give the AI a second chance to detect the intended meaning. However, the illusion of a real conversation breaks if this strategy is used too often.

If Mira still fails to seize the interlocutor's intent, other strategies have to be applied. Mira could take the initiative and might try to guide the conversation back to an easier topic by saying: "Very interesting, but maybe we can talk another time about this topic? By the way, I still need to know when you have time for our meeting". Clearly it is a quite difficult task to keep track of a dynamic conversation and this will require a lot of training data. However, conversational AI cloud services are continuously fed with huge amounts of data which improve the underlying neural networks more and more. Conversational AI will become faster, more flexible and more human like in their response. Even emotion recognition will be improved and could build another resource of information, helping AI to better identify the conversational context and intent [41].

In the end, this process of teaching the AI might help real humans to deeper understand the mechanics and structures of human conversation. The design element outlined above, giving human actors the possibility to "leave a message" to the AI by articulating a free answer or reaction during the scene to which the machine does not react directly, but that serves to build up a neural network enabling the virtual agent to act more freely in future, corresponds to our conception of extending the co-creation to the algorithm. By integrating the data collection in the development process, it will be possible to further discuss the whole design with the groups taking part in the design and evaluation of the prototype. Hopefully it will be possible to reflect on and shape the algorithm in a responsible and critical way.

6 Conclusion

Within this chapter we presented the Virtual Skills Lab project, that has set itself the goal of developing and investigating into the training of social skills through interaction with a conversational AI in an immersive VR environment.

We first discussed why training of social skills is gaining importance, followed by a discussion of VR as a core technology for developing this training. Subsequently we provided an in-depth discussion of challenges that arise when designing such a training, especially how to design an appropriate training environment, an appealing training story and, last but not least, how to deal with technical limitations that current conversational AI still has.

By making our design process, assumptions and decisions transparent we want to provide insights how to design immersive VR trainings for the wicked problem of training social skills.

Acknowledgements We would like to thank the company DOKA for their support and their employees for participating in the co-creation workshops.

This work has been conducted in the project "Virtual Skills Lab" (Grant Nr. 872,573), funded by the Austrian Research Promotion Agency (FFG) in the funding program "Ideen Lab 4.0".

We dedicate this paper to Robert Praxmarer, co-founder of Polycular, who ideated and implemented the Virtual Skills Lab together with us and passed away too early in January 2021.

References

1. Schein, E.: Organizational Culture and Leadership. Wiley, Hoboken (2017)
2. Sackmann, S.: Unternehmenskultur: Erkennen—Entwickeln—Verändern. Springer, Wiesbaden (2017)
3. Fligstein, N., McAdam, D.: A Theory of Fields. Oxford University Press, Oxford (2011)
4. Taylor, J., Van Every, E.: The Emergent Organization: Communication as Its Site and Surface. Erlbaum, Mahwah, NJ (2000)
5. Mead, G.H.: Mind, Self, and Society. The University of Chicago Press, Chicago (1934)
6. Goffman, E.: The Presentation of Self in Everyday Life. Doubleday, New York (1959)
7. Zarifian, P.: Travail et Communication. PUF, Paris (1996)
8. Rullani, E.: Economia Della Conoscenza. Carocci, Roma (2004)
9. Boltanski, L., Chiappello, È.: Le nouvel ésprit du capitalisme. Gallimard, Paris (1999)
10. Cooren, F.: The Organizing Property of Communication. John Benjamins, Amsterdam, New York (2000)
11. Gramaccia, G.: Les actes de langage dans les organisations. L'Harmattan, Paris (2001)
12. Zarifian, P.: A quoi sert le travail? La Dispute, Paris (2003)
13. Kanning, U.: Soziale Kompetenzen fördern. Hogrefe, Stuttgart (2009)
14. Waterworth, J., Riva, G.: Feeling Present in the Physical World and in Computer-Mediated Environments. Palgrave Macmillan, New York (2014)

15. Gillies, M., Pan, X.: Virtual reality for social skills training. In: Hudson, J., Kerton, R. (eds.) Virtual and Augmented Reality to Enhance Learning and Teaching in Higher Education Conference 2018, pp. 83–92. IM Publications Open, Chichester (2019)
16. Pan, X., Hamilton, A.: Why and how to use virtual reality to study human social interaction: the challenges of exploring a new research landscape. Br. J. Psychol. **109**(3), 395–417 (2018)
17. Schmid Mast, M., Kleinlogel, E., Tur, B., Bachmann, M.: The future of interpersonal skills development: immersive virtual reality training with virtual humans. Hum. Resour. Dev. Q. **29**, 125–141 (2018)
18. van Loon, A., Bailenson, J., Zaki, J., Bostick, J., Willer, R.: Virtual reality perspective-taking increases cognitive empathy for specific others. PLOS ONE **13**(8), e0202442 (2018)
19. Bailenson, J.: Experience on Demand. W. W. Norton & Co, New York (2018)
20. Welleck, S., Weston, J., Szlam, A., Cho, K.: Dialogue natural language inference. arXiv:1811.00671 [cs.CL] (2019)
21. See, A., Roller, S., Kiela, D., Weston, J.: What makes a good conversation? How controllable attributes affect human judgments. arXiv:1902.08654 [cs.CL] (2019)
22. Rashkin, H., Smith, E.M., Li, M., Boureau, Y.: Towards empathetic open-domain conversation models: a new benchmark and dataset. arXiv:1811.00207 [cs.CL] (2019)
23. Burkhardt, F., Weiss, B., Eyben, F., Deng, J., Schuller, B.: Detecting vocal irony. In: Rehm, G., Declerck, T. (eds.) Language Technologies for the Challenges of the Digital Age. GSCL 2017. LNCS, vol. 10713, pp. 11–22. Springer, Cham (2018)
24. Mursion Homepage. https://www.mursion.com/services/corporate/leadership/. Last accessed 8 Oct 2021
25. Talespin Homepage. https://www.talespin.com/copilot. Last accessed 8 Oct 2021
26. Strivr Homepage. https://www.talespin.com/copilot. Last accessed 8 Oct 2021
27. Equal Reality Homepage. https://equalreality.com. Last accessed 8 Oct 2021
28. Bodyswaps Homepage. https://bodyswaps.co/. Last accessed 8 Oct 2021
29. https://vhil.stanford.edu/projects/2020/empathy-and-perspective-taking/. Last accessed 8 Oct 2021
30. http://beanotherlab.org/. Last accessed 8 Oct 2021
31. https://www.kalinabertin.com/manic-vr/. Last accessed 8 Oct 2021
32. Gebhard, P., Schneeberger, T., Baur, T., André, E.: MARSSI: model of appraisal, regulation, and social signal interpretation. In: Dastani M., Sukthankar G., André E., Koenig S. (eds.) Proceedings of the 17th International Conference on Autonomous Agents and Multiagent Systems, IFAAMAS, pp. 497–506 (2018)
33. Mori, M., MacDorman, K., Kageki, N.: The uncanny valley. IEEE Robot. Autom. Mag. **19**, 98–100 (2012)
34. Mehl, M., Vazire, S., Ramírez-Esparza, N., Slatcher, R., Pennebaker, J.: Are women really more talkative than men? Science **317**, 82 (2007)
35. Sai, F.Z.: The role of the mother's voice in developing mother's face preference: evidence for intermodal perception at birth. Infant Child Dev. **14**(1), 29–50 (2005)
36. Schulz von Thun, F.: Miteinander reden 1. Störungen und Klärungen. Rowohlt, Reinbek bei Hamburg (1981)
37. Ekman, P., Friesen, W.: Unmasking the Face. Prentice Hall, Englewood Cliffs (1975)
38. Simondon, G.: L'individuation Psychique et Collective. Aubier, Paris (2007)
39. Lawrence, J., Reed, C.: Argument mining. Comput. Linguist. **45**(4), 765–818 (2019)
40. FeldmanHall, O., Shenhav, A.: Resolving uncertainty in a social world. Nat. Human Behav. **3**(4), 426–435 (2019)
41. Hazarika, D., Poria, S., Zimmermann, R., Mihalcea, R.: Conversational transfer learning for emotion recognition. Inf. Fusion **65**(1), 1–12 (2021)

The Brokenness in Our Recommendation Systems: Computational Art for an Ethical Use of A.I.

Giulia Taurino

Abstract Online recommendation systems are information filtering systems that provide users with streams of prioritized content based on expected individual preferences. While they can be of different types—e.g. collaborative, content-based, or hybrid filtering, they typically share the use of machine learning as a type of artificial intelligence able to perform predictions and profile personal taste. Drawing upon previous research in critical algorithm studies, this paper tackles the limitations of predictive content personalization and automated sorting. It does so by taking as a case study the computational art project *This Recommendation System is Broken*, developed in collaboration with metaLAB (at) Harvard as part of *Curatorial A(i)gents* (2020–2022), a series of experiments at the Harvard Art Museums exploring the interplay between A.I. and curatorial practices. While advocating for the ethical design and use of artificial intelligence, I discuss creative coding (Maeda in Creative code, Thames & Hudson Inc., New York, 2004 [1]) as a mode of engagement for artists, designers, media practitioners to take action in the development of context-sensitive algorithms that promote speculative (Dunne and Raby in Speculative everything: design, fiction and social dreaming. The MIT Press, 2013 [2]) design practices and sustainable future-making (Yelavich and Adams in Design as future-making. Bloomsbury Academic, London, GB, 2014 [3]), rather than adopting predictive or statistical models. The art project presented here challenges us to consider the biases of automated decision-making in generating instances of visibility/invisibility on media platforms and other online environments. What we might call "brokenness" is ultimately an attempt to escape the illusory quest for certainty and artificial perfection. Even more so, it is about shaping an ethic of A.I. practice and understanding how information filtering systems are transforming contemporary media cultures.

Keywords Computational art · Algorithms · A.I. ethics · Online platforms · Recommendation systems

G. Taurino (✉)
Experiential AI Postdoctoral Fellow, Northeastern University, Boston, USA
e-mail: g.taurino@northeastern.edu

A. Dingli et al. (eds.), *Disruptive Technologies in Media, Arts and Design*, Lecture Notes in Networks and Systems 382, https://doi.org/10.1007/978-3-030-93780-5_11

1 Introduction

Online recommendation systems are information filtering systems that provide users with streams of prioritized content based on expected individual preferences. While they can be of different types—e.g. collaborative, content-based, or hybrid filtering, they typically share the use of machine learning as a type of artificial intelligence able to perform predictions and profile personal taste. Drawing upon previous research in critical algorithm studies, this paper tackles the limitations of predictive content personalization and automated sorting. It does so by taking as a case study the computational art project *This Recommendation System is Broken*, developed in collaboration with metaLAB (at) Harvard as part of *Curatorial A(i)gents* (2020–2022), a series of experiments exploring the interplay between A.I. and curatorial practices in museums.

By advocating for the ethical design and use of artificial intelligence, I notably discuss creative coding [1] as a mode of engagement for artists, designers, media practitioners to take action in the development of context-sensitive algorithms that promote speculative [2] design approaches and sustainable future-making [3], rather than adopting predictive or statistical models. In line with a shared conversation on media, arts and design, both this article and the project presented here center on the identification, analysis and re-semantization of faults, gaps, errors in A.I.-driven practices for content filtering, thus favoring epistemological ruptures in the understanding of algorithmic technologies and inviting users to think about the social impact of recommendation systems.

First, I will consider the metaphor of brokenness in relation to the study of online platforms and algorithmic media. In a second part, I will contextualize the methodological approach of this paper in the field of critical algorithm studies. Finally, in the third and fourth part, I will provide a series of examples in creative coding applied to digitized art collections, by presenting, among others, a detailed description of the ongoing project *This Recommendation System is Broken*, based on data from the Harvard Art Museums. By leveraging on a conceptual use of programming methods in connection with museums' archives and databases, this computational project recreates a "broken system" that challenges the audience to explore the biases of machine learning in generating instances of visibility on media platforms, while also problematizing pre-defined assumptions in art history. What we might call "brokenness" is therefore about exploring an ethics of algorithms, one that generates awareness on how information filtering systems are transforming the access to digital culture, as well as our very understanding of media and the arts.

2 A Metaphor for the Digital Age

Breaking up, breaking down. Breaking out, breaking through, breaking forth. The metaphor of brokenness is becoming more and more relevant in the study of new media technologies and algorithmic culture. It points at dysfunctionalities in the system, widespread defects that often happen at the operational level of software design, from the selection of training sets for machine learning to the actual application of algorithmic formulae. Concealed or exposed, these defects have major consequences on individual lives as much as on the broader techno-cultural shaping of contemporary societies. Digital models influencing social behaviors through the interaction with technological tools are already embedded in the algorithmic design of online platforms. One of the most discussed applications related to online media involves the large-scale diffusion of algorithmic recommendation systems in the sorting and filtering of content. The relevance of algorithms [4] and their controversy spanned throughout different disciplines and we now witness an unprecedented urgency to create more diverse, inclusive, sustainable practices in algorithmic design.

The concept of a broken machine [5] in algorithms originates from previous research I conducted in the field of media studies and digital ethnography. While researching online platforms, I noticed ongoing conversations pointing at the limitations and imperfections of algorithmic classification. For instance, around 2018, some users started complaining across a variety of online communities about errors found in recommendation systems on streaming services like YouTube, by hinting at the fact they might be broken. Public comments on the web ranged from questions like "Is Recommended Videos currently broken?" and "Is YouTube's recommendation system broken?", to reviews like "This is weird, YouTube must be broken"., or "Honestly, YouTube's recommendation system is broken. None of these are relevant at all to the video."[1] Around the same time, Renee DiResta, research manager at the Stanford Internet Observatory, wrote an article on *Wired* about the need for better recommender systems, acknowledging that recommendation engines are, so to speak, broken. "The recommendation engines we engage with are broken in ways that have grave consequences: amplified conspiracy theories, gamified news, nonsense infiltrating mainstream discourse, misinformed voters. Recommendation engines have become The Great Polarizer" [6].

Since then, scholarships concerned with the ethical issues of using algorithms for content organization on social media and streaming platforms gained more and more resonance. A series of studies on the effects of social media and other online platforms in strengthening racial and gender biases, extremist views, negationism, misinformation appeared on academic journals and publications leading to the theorization of "algorithms of oppressions" [7] and discriminatory operations hidden in the code [8]. The conversation about the risks of new technologies is now entering the popular discourse, raising the need for an inclusive social vision able to promote media and computational literacy. Here I propose to adopt the concept of "brokenness" not only to revert the biases of algorithmic recommendation systems, but also to advance a new media practice based on an ethical and educational scope that combines computational art with creative coding techniques. In this innovative vision, broken machines

---

[1] https://www.reddit.com/r/youtube/comments/8j93pg/honestly_youtubes_recommendation_system_is_broken/

[5] and broken knowledge [9] blend together in a feminist metaphor for the digital age.

As the terminology of brokenness spreads among academic and non-academic communities, questions arise on how we can identify and define a broken recommendation system. Is it a broken code? Is it a broken algorithm? When it comes to software design, a common suggestion is not to leave "broken windows" unrepaired, meaning bad designs, wrong decisions, or simply poor code [10, p. 5]. In Hunt and Thomas' Broken Windows Theory, flaws or imperfections in the code are considered potential hazards leading to further issues, mistakes, irregularities, accelerated software entropy [10, p. 4]. However, fixing the code in more technical terms did not prove to be enough to fix errors passively inherited or actively induced by pre-existing human socio-cultural biases. Once the code is fixed, there is still the risk of relapse into implicit epistemological limits in how we determine such brokenness from a social and ethical standpoint. In the following paragraphs, I will observe what this brokenness represents from a perspective that includes the humanities and the arts in the study of algorithmic technologies.

3 Critical Algorithm Studies

The discipline that studies the social impacts, biases and limitations of algorithms from a critical humanistic standpoint is known as Critical Algorithm Studies. This discipline treats "algorithms" as relevant objects for academic fields beyond computer science, engineering and other scientific domains—to embrace cultural and media studies, but also social sciences. Common topics explored within this field range from the study of relationships between algorithms and society, or between algorithms and cultures, to the analysis of algorithmic governance and policy, up to the design of algorithms and their human component. All these topics rotate around a central question: what are algorithms? In 2017, the scholar Nick Seaver suggested that we should look at algorithms as culture, not just within or in relationship with culture [10]. Instead of offering a "correct" mathematical definition, he proposed to approach algorithms from a pragmatic standpoint, that is to say considering their empirical and practical existence. Within this critical framework, I ask a more specific question: what is the interplay between algorithms—such as those underlying recommender systems, and cultural collections—namely, art collections?

Such a question, along with the project I present in this article, echoes four recurrent subjects in the study of algorithmic practices: media, art, design and A.I. As the paper covers these four domains, a general definition of algorithmic recommendation systems is needed. Online recommendation systems are information filtering systems that provide users with streams of prioritized content based on expected individual preferences. For example, content-based recommendation systems build upon personal taste and the content liked over time to make new suggestions. Or else, collaborative recommendation algorithms sort content based on users sharing similar tastes and preferences. Other algorithms combine both. While they can be of different types—e.g. content-based, collaborative or hybrid filtering, recommendation systems typically share the use of machine learning technologies as forms of artificial intelligence able to perform predictions and profile personal taste. It is widely proven that algorithms for predictive content personalization and automated sorting bear biases and limitations, especially when applied to socio-cultural data. There is a need for defining a new ethic of algorithms. Even more than algorithms as culture, we should start thinking about ethical algorithms and socially-aware algorithmic design [11]. As scholars, researchers, artists, designers and media practitioners, we are called not only to discuss the ethical use of A.I., but also to actively engage in the creation of context-sensitive algorithms that account for the socio-cultural environment that surrounds them.

When I was asked to reflect on AI-driven practices in application to digitized museum collections for the metaLAB's project Curatorial A(i)gents, I started from three main points: first, the code I created had to be context-sensitive, meaning it should account for the institutional context of the collection itself. Moreover, it had to be culture-specific, meaning it should be aware of the cultural context it operates in. Finally, it should be oriented towards practices of future-making [3, 12], instead of focusing on prediction. To this regard, early studies on recommendation systems have shown that the creation of Context-Aware Recommender Systems (CARS)—that is to say, inserting in the machine learning process contextual information and variables (time, location, social context)—allows for a more effective machine learning. The research on this topic includes, among others, a multiyear-long project funded by the European Union and coordinated by the Spanish innovation company Telefonica Investigacion y Desarrollo, between 2011 and 2013. The research project states that:

> Recommender Systems have become essential personalized navigational tools for users to wade through the plethora of online content as they allow users to discover relevant information that they would have never known it existed. In recent years, the importance of this information discovery process as opposed to explicit (keyword-based) search has been emphasized. Current research in Recommender Systems, while taking into account the relation between user and item, often ignores the "context" of the recommendation. We define as "context" any environmental, temporal or otherwise variable that influences a decision a user might make.[2]

Here I propose to bring this framework to the fore, to include a complete revision of what recommender systems are intended to provide in institutional contexts and when applied to cultural objects. Bearing in mind this aim, I started tinkering with exploratory programming [12] and available art datasets. The intent was to investigate possible applications of an algorithmic design that could account for context-awareness and cultural specificity through critical making and research-creation. The result is a research-based art project that dismantles the very principles of algorithmic recommendation, and offers a radical solution.

---

[2] CORDIS—EU research results *Context-Aware Recommender Systems (CARS)* https://cordis.europa.eu/project/id/273739.

4 This Recommendation System Is Broken

The project I present here is part of Curatorial A(i)gents, "a metaLAB series of interactive projects exploring machine learning in, around, and about the Harvard Art Museums collection."[3] I created *This Recommendation System is Broken* as a computational art project that advocates for a socially-aware, context-sensitive, culture-specific, future-making oriented algorithmic design for recommendation systems. It does so by inserting randomization in the selection of content, based on the least viewed works of art in the overall collection. To understand this operation, it is important to understand the institutional history of the dataset used for this project. The Harvard Art Museums comprise three museums: the Fogg Museum, the Busch-Reisinger Museum, and the Arthur M. Sackler Museum. The overall collection has over 200,000 works of art. Only less than 1500 are currently visible in the museums, which is to say less than 1% of the collection is being seen in the present state.

A predictive recommendation algorithm would probably neglect a large part of the collection, and most works of art would remain unseen, invisible. It would rely upon conventional, longstanding, consecrated associations: between artists, art movements, periods, schools, and subject matters. The following project turns this familiar matchmaking practice on its head, seeking to transport the viewer from center to periphery, from the gallery wall to deep in the storage closet, from the highly valued to the undervalued and rarely seen. For instance, imagine you are looking at *The Gare Saint-Lazare: Arrival of a Train*, painted in 1877 and considered to be one of the most iconic Claude Monet's paintings. If you like it, most recommendation algorithms might suggest other impressionist paintings, perhaps located in the same section of the museum. Or maybe, they will forward you to other versions of this painting, at Musee d'Orsay, in Paris, and at the Art Institute of Chicago. Or, again, they will tell you that you might like more works by the world-famous French artist, maybe *Red Boats, Argenteuil*. You are at the center of one of the most popular 19th-century European art movements. But what if artificial intelligence could take you outside the room, far from the center? What if artificial intelligence could take you at the margins, to discover undervalued, hidden artworks?

This recommendation system is, in a sense, broken, which is to say, inaccurate. It does not respond to personal choices; it doesn't collect data; nor does it attempt to interpolate preferences, identify interests or generate predictions. Instead, it makes visible otherwise invisible artifacts in the collection, filtering for objects either rarely accessed or supported by incomplete records. By running the code, the user could end up admiring untitled objects with curious descriptions (Fig. 1), or discovering works of arts from unidentified regions and cultures (Figs. 3 and 4). Among the objects found by this *broken* recommendation system is a "Mural Fragment with Later Overpainting" by an unknown artist (Fig. 2). This fragment dates back to twelfth-thirteenth century's Iran and it is part of the Seljuk-Atabeg period in Persian Art. An impressive work of art, with a rich history, that we probably wouldn't have seen if we had used a typical recommendation system (Fig. 2, 3 and 4).

[3] https://metalabharvard.github.io/projects/curatorial-aigents/.

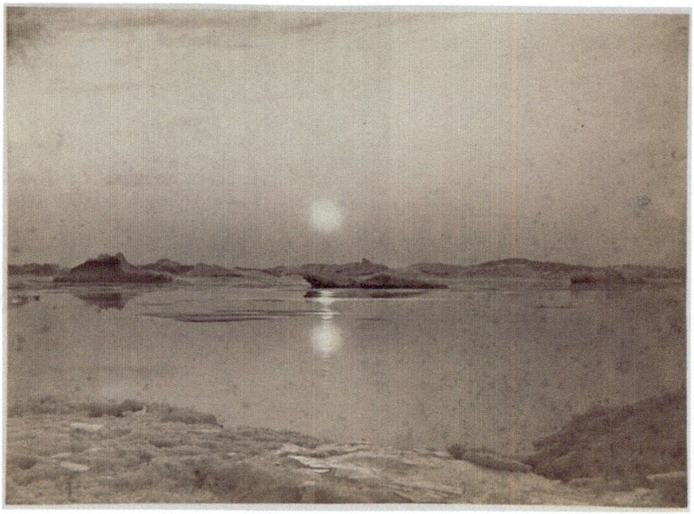

Fig. 1 Author: William Bradford; Title: Untitled; Date: 1869; Courtesy of: Harvard Art Museums/Fogg Museum, Transfer from the Carpenter Center for the Visual Arts, Transferred from Widener Library; © President and Fellows of Harvard College; Link: https://hvrd.art/o/159108

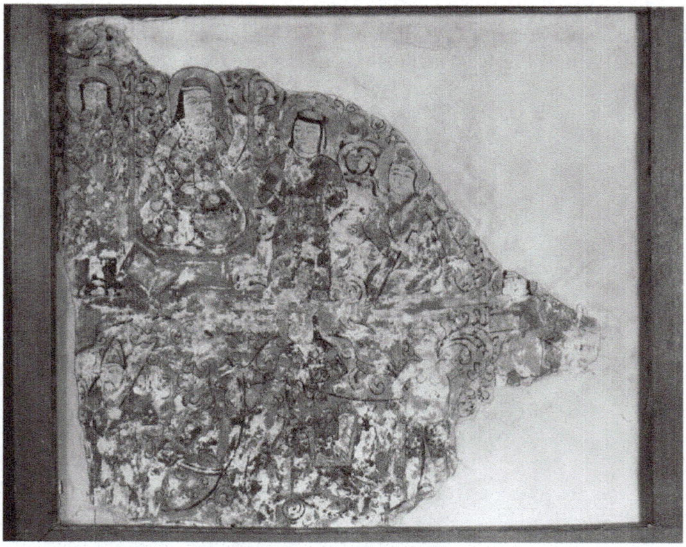

Fig. 2 Author: Unknown Artist; Title: Mural Fragment with Later Overpainting; Date: twelfth-thirteenth century; Courtesy of: Harvard Art Museums/Arthur M. Sackler Museum, Francis H. Burr Memorial Fund; © President and Fellows of Harvard College; Link: https://hvrd.art/o/216793

H/

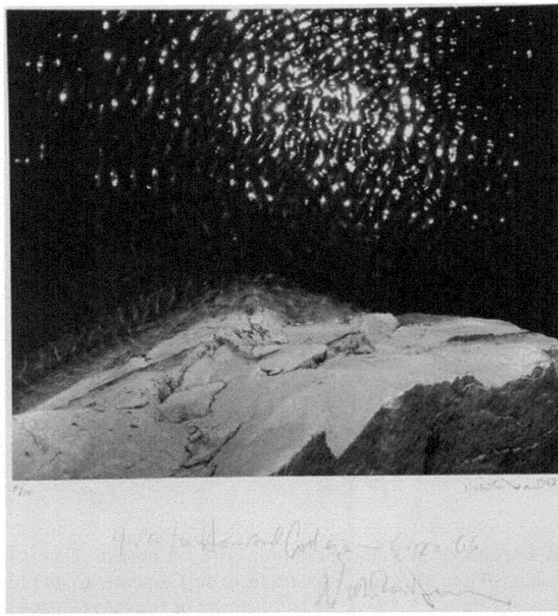

Fig. 3 Author: Nicholas Dean; Title: Untitled (nature abstraction); Date: 1963; Place: Unidentified Region; Courtesy of: Harvard Art Museums/Fogg Museum, Transfer from the Carpenter Center for the Visual Arts © President and Fellows of Harvard College; Link: https://hvrd.art/o/156218

H/

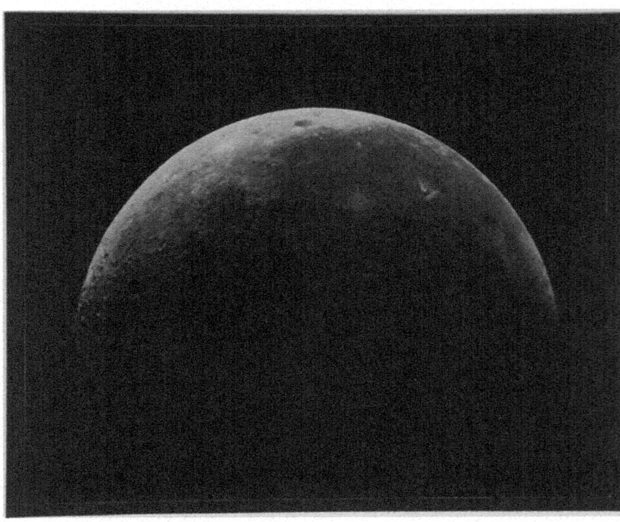

Fig. 4 Author: Unidentified Artist; Title: Untitled (the moon); Date: n.d.; Culture: Unidentified culture; Courtesy of: Harvard Art Museums/Fogg Museum, Transfer from the Carpenter Center for the Visual Arts © President and Fellows of Harvard College; Link: https://hvrd.art/o/159167

Built upon the Harvard Art Museums' dataset, accessible via API, with its set of metadata descriptors and system of categorization, the project employs coding techniques to disrupt norms, to de-center the viewer's perspective, to generate alternative instances of visibility, while reimagining conceptual peripheries within the museum collections and rehabilitating forgotten objects from marginalized art histories. Some amount of randomization is inserted in the code to argue for a new perspective on the concept of brokenness in algorithmic design, one that leads to new discoveries rather than reiterating predetermined expectations. In this continuous process of discovery, the algorithm created for this project acts as a form of epistemic intervention, by changing the very process of knowledge production, stimulating an unexpected order of retrieval and making visible hidden or lesser known works of art.

5 Favoring Epistemological Ruptures in Algorithmic Practice

Beyond more scientific definitions, in popular belief a broken algorithm is perceived as one that fails in performing accurate prediction. And yet, precisely embracing this form of brokenness can lead to new explorations and knowledge, overcoming the limitations of using biased training sets for machine learning. Drawing upon a critical and ethical approach to algorithms and new media practices, the project I developed for the exhibition series *Curatorial A(i)gents* starts from the premise that art should not be handled in the same way as finance or marketing. The creation of algorithms for cultural institutions or media services should not be guided by prediction, popularity or money-making. Finance works at a scale, culture doesn't. Admitting a framework where art and cultural institutions do not need to function at scale is a fundamental step to decolonize museums and acknowledge that art collections have very specific identities, histories, traditions that cannot be treated in a systematic, standardized, pre-defined way. We need diversity in algorithms, meaning algorithms that are tailored for each cultural institution.

The Harvard Art Museums in Cambridge, MA might need a different algorithm than the Museum of Modern Art in New York or the National Museum of Anthropology in Mexico City. Implementing creative coding methods in the arts and cultural heritage sectors offers a viable solution for making algorithmic design more inclusive and diverse. Some projects have already been developed in this direction, such as *Art I Don't Like* by Sarah Frost and Manu Mathew Thomas based at the Creative Coding Lab at the University of California, Santa Cruz. *Art I Don't Like* is an "anti-recommender" system, where the algorithm filters works of art with a low similarity score, finally showing a work that users might not like. This is just one example of prototypes that integrate neural network technologies with creative coding for managing art and media collections. Similar experimental projects include Google Arts and Culture's *X Degrees of Separation* (https://artsexperiments.withgoogle.com/

xdegrees/), which reconstructs different paths and degrees of connection between two artworks, or *MosAIc* (https://microsoft.github.io/art/app), a software developed by MIT and Microsoft researchers with the aim of finding hidden associations between analogous art objects, using data from The Rijksmuseum in Amsterdam and The Metropolitan Museum of Art in New York [13].

Other projects focusing on AI and museums use algorithms as curators, in order to create a web-based, data-based selection of content that goes outside of the often partial human perspective. The Museum of Digital Art in Zurich, Switzerland for instance uses an algorithm called Hal 101 to compare past patterns, internet searches, available data to select and organize artists and content for future exhibitions. Artists are then responsible, in a second phase, for curating their own exhibition. This is meant to avoid biases in selecting artists' nationality, age, gender or financial factors. Another example is *Recognition*, winner of IK Prize 2016 for digital innovation, "an artificial intelligence program that compares up-to-the-minute photojournalism with British art from the Tate collection" using computer vision. (http://recognition.tate.org.uk/archive/info) Over the course of three months, Recognition's algorithm created a virtual gallery by "searching through Tate's collection of British art and archive material online, comparing artworks with news images from Reuters based on visual and thematic similarities". (*ibidem*) Such projects use machine learning to generate unforeseen, unpredictable connections across history, geography, culture.

Even before these more recent examples, an attempt to revolutionize recommendation systems appeared in 2006 under the name of *Unsuggester*. The *Unsuggester* was introduced as "the worst recommendation system ever devised",[4] since it recommended books using the format "if you have read this, you are unlikely to read this". In her book *The Participatory Museum*, Nina Simon asks us to imagine a similar principle applied to museum visits. "People might be intrigued to learn that 'if you always visit the mummies, you may never have explored the fish tanks'. Recommendation systems must meaningfully respond to users' profiles, but they don't have to be optimized solely to provide people with more of what they already like" [14, p. 62].

6 Conclusion

Reasoning on *This Recommendation System is Broken* and other projects, this article sheds a light on the urgency to prevent a small set of algorithms to define all these interrelated media and cultural processes we live in every day. It also opens up to a discussion on governance, privacy and the ways personal data are handled. By addressing the complex intertwining between arts and artificial intelligence, my project for metaLAB (at) Harvard invites researchers, practitioners, users to question the need for prediction and personalization when it comes to accessing cultural and media content. By doing so, it advances alternative possibilities for algorithmic

[4] https://blog.librarything.com/main/2006/11/booksuggester-and-unsuggester/.

design, fostering new epistemological directions and explorations of our own cultural biases. As we approach a seemingly perfect, non-human sense of artificial intelligence, the risk is to become unable to understand the complexity of it and its inner brokenness. Purposely breaking recommendation systems means opening up a way for creating our own human-oriented algorithms.

Media or computational art can be adopted as a means to problematize and take a critical stance towards the gaps, biases, errors in most common algorithmic practices. The epistemological shift induced by the use of creative coding operates on many levels. On the one hand, it fosters coders to engage into a cross-disciplinary activity, where the functional aspect is connected to a creativeintent, whether in the form of art or activism. Scholars and practitioners from different disciplines meet in this creative space in which exploratory programming, computational art, media design come together to interrogate how new technologies and related practices are affecting societies and cultures. Initiating an interdisciplinary conversation on techno-culture help diminish the silos of research, while also generating a more diverse environment for the development and application of algorithms. On the other hand, creative coding projects are often intended to facilitate a certain level of engagement and interaction from the users, which are invited to take a more active and aware mode to computational technologies. This is part of a pedagogical component of creative coding, which was already discussed in Maeda's groundbreaking work, *Creative Code. Aesthetics + Computation* [1].

The importance of computational literacy lies in bridging the chasm between computer science and social sciences as much as in breaking up with traditional scientific practices in internet research that ended up reinforcing socio-cultural discrimination. Against fixed conceptions of knowledge based on epistemological continuity, "the problem of scientific knowledge must be posed in terms of obstacles" [15], thus making breaks, ruptures, disruptions part of a natural process in the advancement of scientific and social research. Much like other creative coding projects, by privileging media art practice over a more rigorous, technical approach, the art project presented here aims at redefining the use of computational tools in cultural institutions as antidotes by which further clicks reveal ever less familiar artists, defined objects, and identifiable regions. "Brokenness" is a no less relative concept than wholeness. But here it is featured rather than concealed. The term becomes a window into the instrumental ethics of algorithms, one that observes how information filtering systems shape culture by re-inscribing what is always already preferred.

References

1. Maeda, J.: Creative Code. Thames & Hudson Inc., New York (2004)
2. Dunne, A., Raby, F.: Speculative Everything: Design, Fiction and Social Dreaming. The MIT Press (2013)
3. Yelavich, S., Adams, B. (eds.): Design as Future-Making. Bloomsbury Academic, London, GB (2014)

4. Gillespie, T.: Keyword: algorithm. In: Peters, B. (ed.) Digital Keywords: A Vocabulary of Information Society and Culture, Princeton University Press (2016)
5. Sharma, S.: A manifesto for the broken machine. Camera Obscura **35**(2(104)), 171–179 (2020)
6. DiResta, R.: Up next: a better recommendation system. Wired (2018). https://www.wired.com/story/creating-ethical-recommendation-engines/
7. Noble, S.: Algorithms of Oppression: How Search Engines Reinforce Racism. New York University Press, New York, NY, US (2018)
8. Benjamin, R.: Race After Technology: Abolitionist Tools for the New Jim Code. John Wiley & Sons (2019)
9. Marchessault, J.: Marshall McLuhan: Cosmic Media. SAGE (2005)
10. Hunt, A., Thomas, D.: The Pragmatic Programmer. Addison-Wesley (2000)
11. Seaver, N.: Algorithms as culture: some tactics for the ethnography of algorithmic systems. Big Data Soc **4**(2) (2017)
12. Kearns, M., Roth, A.: The Ethical Algorithm: The Science of Socially Aware Algorithm Design. UK, Oxford University Press, Oxford (2019)
13. Maeda, J.: Design By Numbers. The MIT Press, Cambridge, MA (1999)
14. Hamilton, M., Fu, S., Freeman, W.T., Lu, M.: Conditional Image Retrieval (2020). arXiv:2007.07177
15. Bachelard, G.: The Formation of the Scientific Mind: A Contribution to a Psychoanalysis of Objective Knowledge. Beacon Press (1986 [1938])
16. Simon, N.: The Participatory Museum. Museum 2.0, Santa Cruz, CA (2010)

Exploring Reinforcement Learning: A Case Study Applied to the Popular Snake Game

Russell Sammut Bonnici, Chantelle Saliba, Giulia Elena Caligari, and Mark Bugeja

Abstract Reinforcement Learning is a machine learning approach in which an agent interacts with their environment to gather information, and make an informed decision based on the accumulated information. In this research, we investigate the applicability of various reinforcement learning techniques for Snake, a video game popular on the Nokia 3310 mobile phone. Q-Learning (Quality-Learning), SARSA (State-Action Reward State-Action) and PPO (Proximal Policy Optimization), were implemented and evaluated for Snake. Q-Learning and SARSA did not generate optimal results due to the large environment of the game. Meanwhile, PPO was implemented with three varying approaches for input; a vector, CNN and raycasting based approach. PPO, in conjunction with raycasting, resulted in the best performance, with the snake agent learning for both collecting food and avoiding obstacles. Furthermore, A* Pathfinding was tested and it achieved a performance better than Q-Learning and SARSA but not better than PPO as it was less adaptable to large environments. In the future, agents in large dynamic game environments, may benefit further from utilizing PPO.

Keywords Reinforcement learning · Q-learning · SARSA · PPO

1 Introduction

1.1 Aims and Objectives

To gather an understanding of performant reinforcement learning techniques for snake, notable algorithms are to be implemented and compared in agent behaviour. An understanding will be reached through the following objectives:

R. S. Bonnici (✉) · C. Saliba · G. E. Caligari · M. Bugeja
University of Malta, Msida MSD 2080, Malta
e-mail: ict@um.edu.mt
URL: https://www.um.edu.mt/ict/

A. Dingli et al. (eds.), *Disruptive Technologies in Media, Arts and Design*, Lecture Notes in Networks and Systems 382, https://doi.org/10.1007/978-3-030-93780-5_12

1. Implement a variation of Snake that is similar to the original game but also adapted for an AI agent to be able to take over and play.
2. Introduce AI techniques such as A* Pathfinding and see the effectiveness of such a technique.
3. Analyse the Q-Learning Algorithm and evaluate its performance over different rewards.
4. Implement the Reinforcement Algorithm SARSA and evaluate its performance over different hyper-parameters and rewards.
5. Implement the PPO algorithm by making use of ML-Agents and analyse the different approaches to which this algorithm can be applied.
6. Compare the results generated and distinguish the best reinforcement learning technique from what was implemented and evaluated.

1.2 Snake

Snake [1] is a game originally developed as "Blockade" in 1976 by the Gremlin Industries. This game is well known for its release on the Nokia 3310. Here, the goal is to eat as much food as possible, while avoiding collisions with the borders or itself. The snake will get longer with each eaten food. Variations of the game include the speed of the snake increasing as the snake gets longer or the snake's ability to pass through the borders and emerging from the other side of the game screen.

1.3 Game Mechanics

Similar to the original, the game is single-player. The player needs to eat as much food as possible without crashing into the boundaries or its own body. Collision will result in the termination of the game. A score is keeping track of the number of eaten food objects. The game does not have a termination condition and will only end when a collision occurs.

Movement. At the beginning of the game, the snake is initialized at the centre of the screen moving downwards. The player can move in all four directions, corresponding to the arrow keys of the keyboard or the WASD keys. The snake's movement is achieved by moving the head in the new position and relocating the last body part of the snake to the empty position generated by the movement of the head. The sole limitation of its movement is the fact that the snake is not able to move in the opposite direction of its movement. To display the snake's rotations, each body part has a graphics component with five possible sprites, corresponding to each rotation of the snake's body. For example, when the snake goes up and then right, at the point where the snake turned right, the body part made visible at that point would be that of an upper left corner body piece.

Food. The food is spawned randomly within the boundaries of the game scene. When the food is eaten, it will be re-spawned to another random position. Eating the food will cause the snake's size to increase by one unit. This becomes visual after the next movement of the snake. When the snake moves forward one unit, the last body part is not deleted for that action, thus increasing the body by one unit. Note that in this variation the speed of the snake's movement is not increasing as the snake eats more food.

1.4 Artwork and Theme

The theme adopted for this variation was that of the Chinese Dragon. The name we appointed was Zhulong Rush, inspired by Zhulong: the red solar dragon and god in Chinese Mythology. Original inspired artwork for the dragon's head and body were created to fit the theme, as can be seen in Figs. 1 and 2. A fireball was designed as a replacement to the food collectable as seen in Fig. 3.

2 Background Research

Reinforcement Learning [2–4] is a field in machine learning in which the agent depends on the interaction with the environment and other agents to choose their next action. The goal of the agent is to maximize its rewards. Essentially, a Reinforcement Model is made up of three main components:

1. Environment states
2. Set of actions that an agent can perform
3. Set of reinforcement signals or rewards

The agent will interact with the environment through a set of actions, and the goal of the agent is to find a policy that maps the states to the action that would maximise the reward.

Reinforcement-learning problems can be solved using two main strategies:

1. Searching in the space of behaviours to find a behaviour that performs well in the environment.
2. Make use of statistical techniques and dynamic programming to estimate the success of an action based on a specific state.

One of the fields in which Reinforcement Learning has been introduced is that of gaming and video games [6]. In this field, the reinforcement agents managed to exceed human performance in games such as Go and Atari. In 2017, Silver et al. [7] implemented table rasa reinforcement learning that managed to surpass the superhuman level of play within 24 h in the games of chess, shogi and Go. Reinforcement learning was implemented on the Atari game by Mnih et al. [8] by making use a convolutional neural network trained with a variant of Q-learning (Fig. 4).

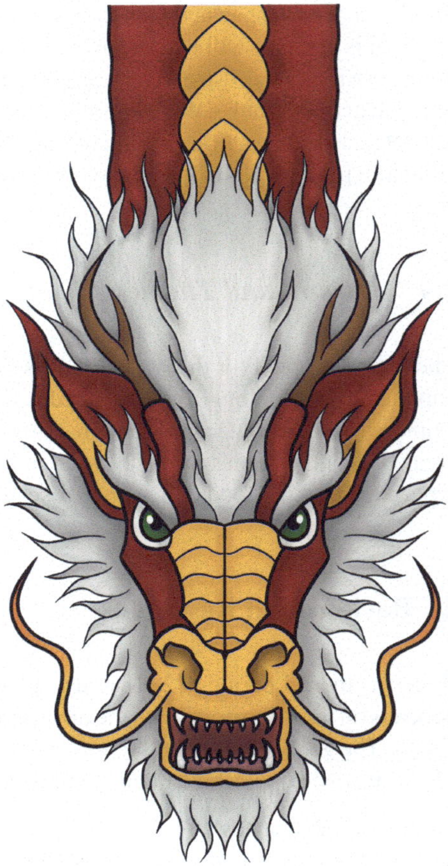

Fig. 1 The Dragon's Head, inspired by Zhulong

Reinforcement Learning was also implemented for Snake by Ma et al. [9] in 2016. In this research, Ma et al. compared the performances of Q-Learning and SARSA reinforcement algorithms. It was concluded that the SARSA algorithm outperformed the Q-Learning algorithm after a long training period. In addition, it was also noted that the Q-Learning algorithm gave unstable results. Furthermore, both algorithms were compared to an optimal solution. From this experiment, they concluded that neither algorithm performed as well as the optimal solution.

Wei et al. [10] in 2018, implemented Snake for Deep Reinforcement Learning. In this research study, the researchers made use of the deep Q-learning network, making use of screenshots of the game to gather information such as the direction and the position of the snake and output an estimated Q-value for each action. A reward mechanism was designed to solve delayed reward issues typically associated with deep Q-learning. Moreover, a training gap strategy and a dual experience replay method were implemented to improve the training efficiency. The system proposed by the researchers made use of a Convolution Neural Network with three convolutional layers and a fully-connected layer.

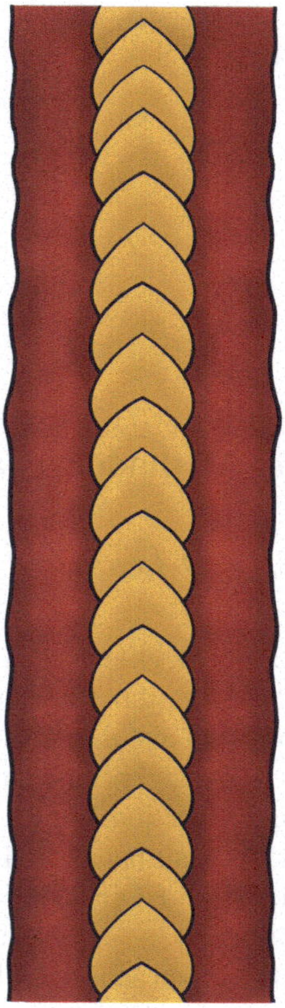

Fig. 2 The Dragon's Body, inspired by Zhulong.

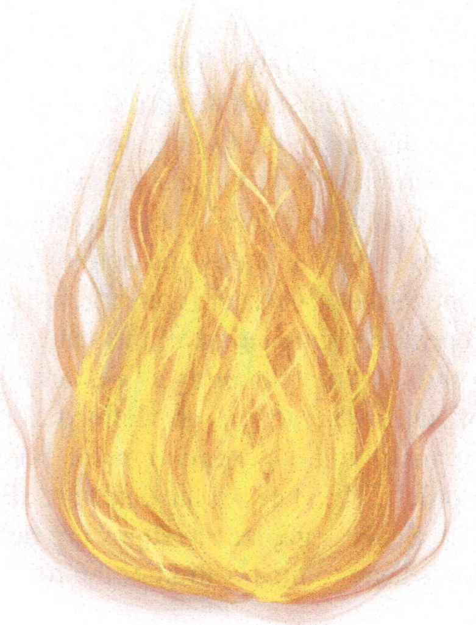

Fig. 3 The Fireball as the new food.

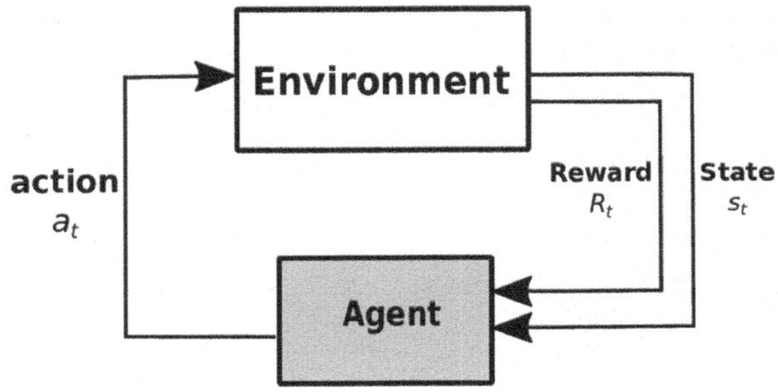

Fig. 4 Reinforcement Learning [5]

3 Methodology and Evaluation

3.1 Q-Learning

Concept. Q-learning [11], developed by Watkins in 1989, is a Reinforcement Learning Algorithm that makes use of the Markov decision process to estimate the environment and model the agent's action based on that environment.

This algorithm is made up of three main components: the environment state, the agent's action and the reward. A Q-table entry is generated for each possible state-action pair based on the reward function [12]. A reward is defined as the goal of the agent, and doing a positive action towards the goal would generate a positive reward. Similarly, a negative action would contribute to a negative reward. An agent always chooses an action that has the maximum reward obtained from its environment stored in the Q-table.

This algorithm needs to make use of a strategy to balance exploration and exploitation. The agent is required to explore the environment and its surroundings in order to generate the Q-Table. Therefore a number of random actions need to be taken for the agent to understand its surroundings. The agent will then start making use of the values achieved in the Q-table to exploit the environment.

The problem with Q-Learning is that for it to be useful, optimal Q-values need to be found. A number of factors contribute to this, such as rewards, number of iterations, possible states and actions and the learning rate. Watkins et al. in 1992 [13] shows that Q-learning converges to the optimal Q-values as long as all the actions are repeatedly sampled in all the states.

Approach Implementing a Q-Learning Algorithm requires a set of actions and a set of states. The actions are defined as all the possible movements that an agent is able to make within the environment. For a Snake's game, these actions are up, down, right or left. The states of the game are all the possible states within the environment. For the Snake's game, the states are defined as the relation of the position of the dragon's head to the position of the food, and it is defined as:

$$(\Delta x, \Delta y) \tag{1}$$

Table 1 Q-Table

	Up	Down	Right	Left
(0, 0)				
(0, 1)				
...				

This shows the difference in the x value and variance of the y value. In addition, the direction is also being taking into consideration. This is being achieved by making use of the addition sign and the negation sign. The addition and the negation signs show the direction that the dragon is required to move to. The addition sign indicates that the snake is needed to move to the right or up while the negation sign shows that the snake is required to move to the left or down. For example, a state of $(-6, -7)$ shows that the snake needs to move six steps to the left and seven steps down in order to eat the food and achieve a reward.

A Q-Table, as shown in Table 1 is initialized for each possible state and action. The number of states is determined by the dimensions of the game. As the game's dimension is increased so are the possible states. The Q-Table was represented as a dictionary to which the keys were the states, and the actions were the values.

Every action that the agent takes will be updated in the Q-Table with the following equation:

$$Q(s_t, a_t) = (1 - \alpha)Q(s_t, a_t) + \alpha(r_{t+1} + \gamma(max\,Q(s_{t+1}, a)) \tag{2}$$

where α is the learning rate, and γ is the discount factor. As one can interpret from this equation, the algorithm makes use of the current Q-value and adjusts this value by calculating the cumulative reward that will be achieved from entering the next state. The learning rate α determines the rate at which the agent would be learning while the discount factor γ determines the importance of future rewards.

An important factor when it comes to Q-Learning is the reward. The agent in the environment is constantly trying to achieve the highest reward and selecting the action that is most likely to maximise the reward. In the game of Snake, there is only one factor to which the agent will receive an award to and this is when the snake finds and eats the food. On the other hand, the agent is to receive a negative reward when the snake hits the boundaries or when the snake hits itself. From experimentation, it was concluded that the reward was more effective when the positive reward was greater than the negative reward. Therefore a positive reward of +5 is given whenever the snake finds and eats the food while a negative reward of -1 is given whenever the snake hits a boundary or itself.

The Q-learning algorithm also makes use of the balance between exploration and exploitation. This is determined by the ε, which is a value between 0 and 1. At the beginning of the game, the agent is required to explore the environment and populate the Q-table with trial and error attempts. As the game progresses, the agent will start to make more informed decisions based on the values in the Q-Table (Exploitation). The ε rate will be decreasing with the number of episodes with a predefined decay (Table 2).

Table 2 Q-Learning hyper-parameters

	Value
Learning rate	0.001
Discount	0.95
Epilson	1.0
Minimum epilson	0.005
Epilson decay	0.99993

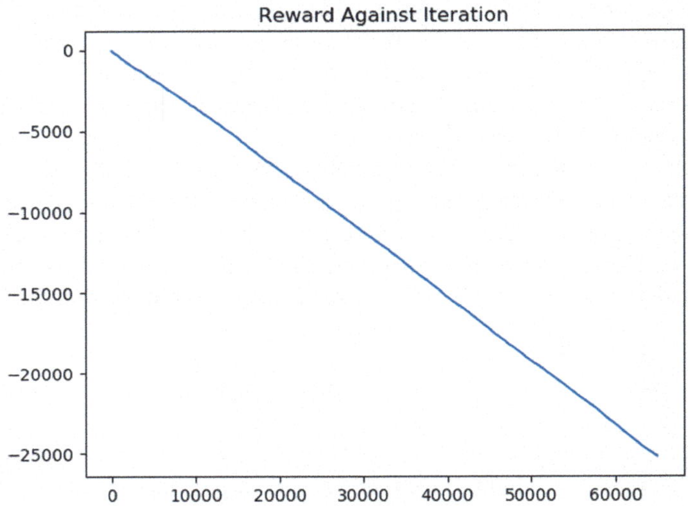

Fig. 5 A graph of cumulative reward against episode

Experimentation. To find the optimal Q-values, the hyper-parameters and the rewards were changed and tested for their improvement of the model.

The first rewards that will be analysed are that of +1 for eating the food and that −1 for hitting the boundary walls or itself. For this experimentation, the agent was left to train for 65,000 episodes. An episode is defined as a game from the agent's initialization till the agent dies and therefore loses the game.

As can be shown, in Fig. 5, the cumulative rewards became more and more negative with every episode. This implied that the agent was dying more often than it was eating the food, and therefore he was receiving more negative rewards than positive rewards accumulating to an average of −25,000 negative rewards over 65,000 iterations. It is important to note that due to the nature of the states, the agent is not receiving any information about the location of the bounding walls or the location of the body. This could have been one of the reasons why the cumulative reward kept getting more and more negative over the number of episodes.

Figure 6 shows the average score every 100 games over the number of episodes. From this graph, one can analyse two important factors that give information about

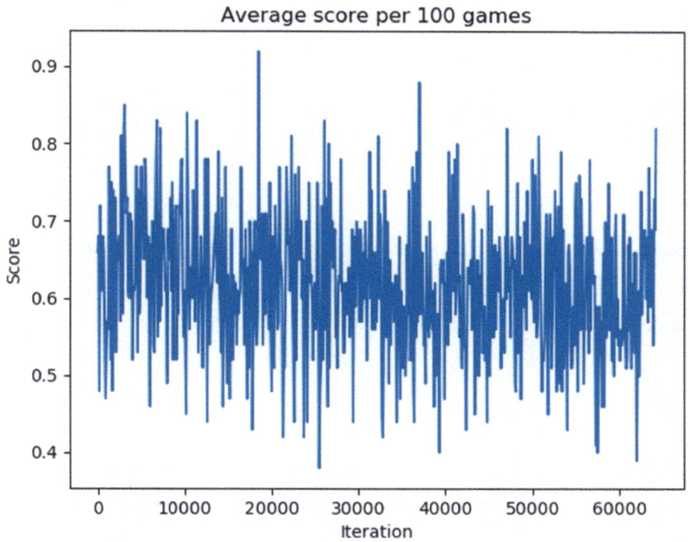

Fig. 6 A graph of Average Score per 100 games against episodes

the current state of the training. The first important factor is that the average score remained approximately the same over the course of the game. Therefore, it was a smooth transition from exploring the environment and exploiting the environment. However, this also shows that there were not many improvements since the agent was not eating more food per episode, and therefore the agent still required more training.

Having a look at the Q-table and the performance of the game at maximum training time, one can note, that the agent managed to learn to get the food when the food was close to the agent. However, when the food was further way, the agent almost never managed to find the food.

The algorithm was also analysed for a reward of +5 for eating the food and a reward of -1 for hitting the boundaries and itself. This analysis was performed over 10,000 iterations.

Figure 7 shows the cumulative reward against the number of episodes. As one can see, since the positive rewards had a higher value, the cumulative reward starting to increase with the number of episodes. Figure 8 shows the average score every 100 games over the number of episodes. From this graph, one can analyse that at around 2000 iterations the average score per 100 games was low; however, this continued to increase over the iterations, reaching a peak at approximately episode 8000.

The different approaches were then compared over a 10000 iterations while retaining the same hyper-parameters to analyse the effect of rewards on the system. Figure 9 shows the average score per 100 episodes over 10000 iterations of the two different reward systems. It was concluded, that the Q-Learning Algorithm managed to perform better when a higher positive reward was given.

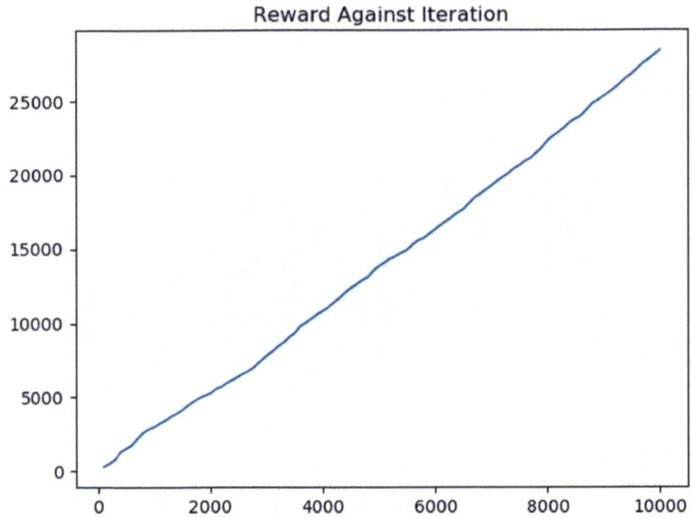

Fig. 7 A graph of cumulative reward against episode

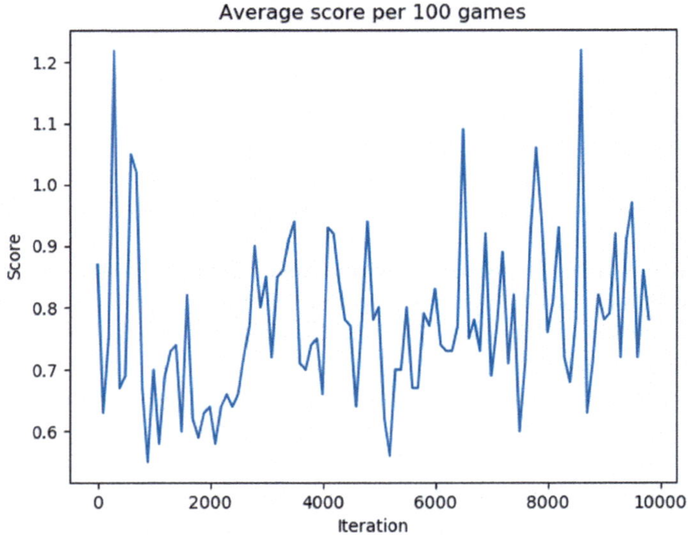

Fig. 8 A graph of average score per 100 games against episodes

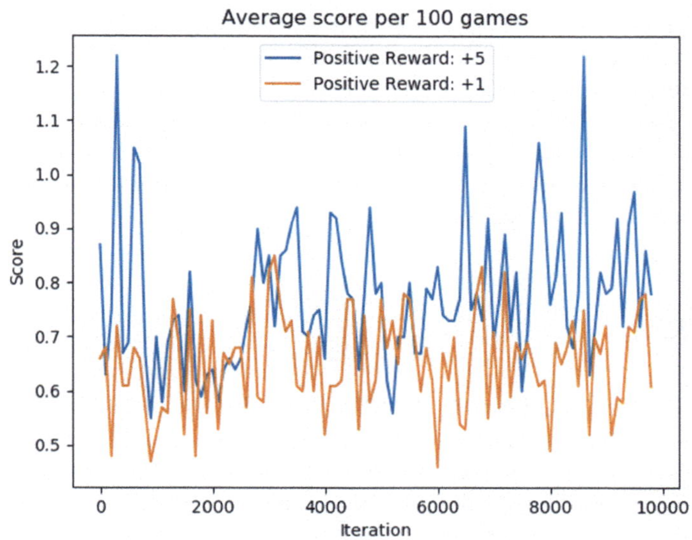

Fig. 9 Comparison of different rewards given

Evaluation. One can conclude that the Q-Learning algorithm was not as successful as it was hoped to be. However, results generated when the positive reward was increased seemed more promising. A number of factors could have contributed to the poor performance of the algorithm.

First of all, as was said before the agent was only receiving direct information about the position of the food with respect to the position of the head. The Q-table did not contain any information regarding the boundary walls and the location of the dragon's body. These were left for the agent to gather knowledge about explicitly.

In addition, the agent was not getting a negative reward for every step that it was taking during the duration of the episode that did not result in a positive reward. Therefore, the agent was not learning to take the shortest path towards the food with a chance of causing uncertainty in the long term because the food could be reached using different paths.

Furthermore, the Q-table is made up of a sizable number of states. This, therefore, does not make it feasible. Currently, the algorithm is being trained on a screen of 20 by 16 Unity units. Therefore the total number of possible states exceeds 500 since we are also taking account the direction (up, down, right and left) of the food to the snake. For each state, there are four possible actions which create a Q-Table of at least 2000 viable cells. As stated by Watkins [13], for Q-Learning to converge, all the actions should be repeatedly sampled in all the states. Therefore, a long training time is required for convergence.

Moreover, Q-Learning also takes a relatively long time to train. Around 50,000 episodes take 10 h of training. This made it computationally challenging to test the performance of different hyper-parameters and different rewards to find the optimal Q-values.

A factor highlighted about Q-Learning by Ma et al. [9] is the fact that Q-Learning could be unpredictable. This was also a fact that was distinguished during the training. Re-training the algorithm with the same hyper-parameters and rewards could sometimes generate different results. This is due to the exploration component of the algorithm that makes use of random actions. Learning by trial and error until exploitation has an element of luck and the agent could get lucky and start finding more food and therefore learn quicker.

3.2 SARSA

Concept. Similarly to Q-Learning, SARSA also makes use of the Markov decision-making process, due to the fact that to find the optimal action, it continuously iterates within its environment. However, as opposed to the Q-Learning algorithm, SARSA is an on-policy algorithm, resulting in the next action to be taken decided upon by a specific policy, which in this case is the current action being taken [14]. SARSA allows for a future reward to be estimated given a state and its associated action, knowing the current policy.

SARSA is an acronym for state, action, reward, state, action, where the first state and action are the current ones being taken, the reward is the observed reward, and the following state and action are the next state in question as well as the action achieved from a greedy policy [15]. The greedy policy is used as in Q-Learning so as to achieve a balance between exploration and exploitation. SARSA also makes use of a learning rate value and discount factor, allowing for learning to occur quickly and giving more importance to current rewards, respectively.

SARSA is seen to perform better in cases where large negative rewards are concerned [16]. This is due to the fact that the SARSA algorithm is able to learn to avoid these negative actions and in turn, learn a policy more time efficiently. This, therefore, makes SARSA a better choice over Q-Learning in situations where potential harm may occur, and so negative outcomes should be avoided as much as possible, as well as require rapid policy learning.

Approach. The SARSA algorithm requires that on startup, all necessary Q values are initialized. There must be a Q value for every state and action pair possible within the environment, and each one is initialized arbitrarily. Therefore, unlike Q-Learning which adds to the Q-Table as the game is played, in SARSA the Q-Table's size and dimensions are established from the very start of the game, however still have the format as seen in 1. For this to be done, a dictionary class was used, representing the Q-Table, where the key was the state, and the value was an array of all possible actions. The state, similarly to the code implemented for Q-learning, contained the x coordinate and y coordinate for the distance between the snake's head and its food, either positive or negative so as to indicate the direction needed to travel. The array of actions was of size four, representing the four possible actions of the snake, to move up, down, left, or right.

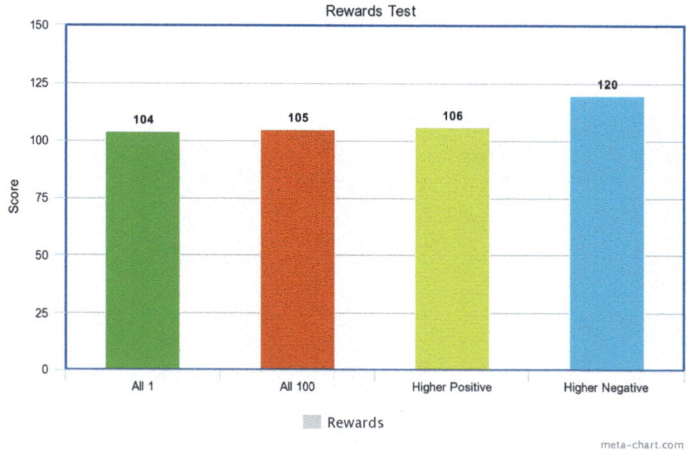

Fig. 10 A graph of scores per reward values

For each time the snake initialized a fresh, the current state is found in respective to the randomly spawning food. The action based on that state is then chosen by means of a policy. This policy allows for a number to be generated, and if within a certain epsilon rate, a greedy policy is taken, however, if the number is within a small percentage that is not within that rate, then a random action is taken on that state, allowing for some exploration to still be achieved. The reward at this point is then observed. The next state is then found along with its associated next action in the same manner as previously discussed.

$$Q(s,a) = Q(s,a) + \alpha(r + \gamma Q(s',a') - Q(s,a)) \tag{3}$$

Equation 3 is then applied so as to update the respective Q value within the table. Where s and a are the current state and action, r is the resultant award, and s' and a' are the next state and action pair. Alpha represents the learning rate and gamma the predetermined discount. The Q value for the current policy is therefore updated with every movement carried out. The current state and action are then updated to equal the next policy.

As in Q-Learning, the reward is increased when the snake collides with the food and is decreased on contact with walls or itself. However, as previously mentioned [16], SARSA works more effectively when concerned with large negative reward values as opposed to the focus being given on the positive rewards. Therefore the rewards were altered as such.

Experimentation. The hyper-parameters, along with the reward values, were experimented with during testing so that they achieve the most significant Q value.

Taking consistent values for the hyper-parameters throughout for fair testing, and an untrained model, different values for the rewards were tested with 1000 iterations each, and their results are seen in Fig. 10.

Table 3 SARSA hyper-parameters testing

	Value	Score
Learning Rate1	0.1	93
Learning Rate2	0.5	72
Learning Rate3	0.9	120
Discount1	0.1	104
Discount2	0.5	98
Discount3	0.9	120

The first case tested out contains equal values for both positive and negative rewards; however, this value is minimal, as it is that of 1 and -1, respectively. The second test similarly contained equal positive and negative reward values; however, this value was raised to see the effect the size of the value has over the result irrespective of ratio. The third test saw the resultant reward obtained when using a greater positive reward in respect to the negative one, and the final test made use of the opposite, where the negative reward was set to be much higher than the positive one.

From the results seen in Fig. 10, it can be noted that there was little to no difference between the results achieved within the first two tests, showing that the ratio between the results has a much larger effect on the outcome as opposed to the size of the value itself. However, the larger values for the reward still gave a slight increase in score value. It can also be noted that a considerable positive value along with a small negative reward value barely alters the results as compared to when the two values were equal. The only rewards combination that seems to cause a significant improvement over the other results is when a large negative reward is used as well as a small positive reward, and this, therefore, confirms the literature discussed earlier.

Due to the previous testing, the rewards used for the next tests were set to a contain a sizeable negative reward when the snake collides with what it is not supposed to and a smaller positive reward for when it finds the desired food. Next, different values for the learning rate and discount value were tested. Each value was verified independently with the learning rate unchanged between tests to find the discount value and discount value consistent between tests for the learning rate. The values were chosen to be over a large scale so as to test a variety of ranges with roughly equal distribution. The values were once again tested on an untrained model for 1000 iterations.

As seen in the Table 3 results , when testing the learning rate, the score obtained is highest with a high value of around 0.9, lower values are seen to impair testing. This could be due to the fact that since the learning rate determines the rate at which the model learns, a higher value results in a model learning quicker, resulting in a higher score. It can also be seen that similarly, the highest score is achieved when the discount score is around 0.9. The discount indicates the importance given to future rewards, and therefore in this situation, a higher needed discount rate is required so as to find the path towards the positive reward of the food.

While conducting continuous long term training, the epsilon value was altered so as to allow for more exploration and less exploitation so as to allow for a higher chance for unseen states to be encountered.

Evaluation. The SARSA algorithm requires a large number of iterations for the system to learn. Therefore a save and load function was added so that training could occur over multiple points in time without progress being lost. This is done by the Q values table being loaded up from where it left off once the program is started up again. The table is saved each time the snake dies.

The SARSA algorithm, however, does not work as it is fully intended, and this is due to a number of contributing factors similarly to as seen in Q-learning.

As mentioned in the evaluation of Q-Learning, the Q-Table only contains the distance between the head of the snake and the food. This proves to cause some problems as it does not contains any information about the snake's distance from the walls and possible snake body parts. This results in the snake colliding with the wall on several occasions even if it has been in that position before, due to the fact that the distance from the food may be different whilst the distance from the wall is not, resulting in the Q Table stating that, that state has not been encountered yet.

The SARSA algorithm also requires the size of the Q-Table to be preset from the beginning. This results in a great memory being taken up as there exist over 2500 possible states. Also, whereas with Q-Learning, the states are added over time, with SARSA this large table is always present.

The time needed to train the model is also considerable. With around 100,000 iterations of training, over countless hours, the Q-Table is still only about half way full, with a number of states yet not encountered by the model. Figure 11 shows a small portion of the values within the trained Q-Table. Each entry shows the state, action from that state and the respective Q value. As seen from the image, specific values still contain the Q value 0, which was the arbitrary value assigned for all Q-values upon creation of the Q table. This indicates that that policy has yet to be encountered by the learner even after the extensive time of training.

3.3 PPO

Concept. Proximal Policy Optimization [17], engineered by Schulman et al. in 2017, is a reinforcement learning algorithm that outperforms other online policy gradient methods in sample complexity, simplicity and wall-time. Proximal Policy Optimization (PPO) uses a neural network for approximating the ideal function that maps the agent's current observations to the best possible action the agent can take. In order to tweak the weights of the neural network to achieve the desired function, a cost function is applied along with gradient descent. At each step, PPO updates the neural network to minimize the cost function while simultaneously ensuring that the standard deviation from a previous policy is relatively small, avoiding overfitting.

$$L^{CLIP}(0) = \hat{E}_t[min(r_t(\theta)\hat{A}_t), clip(r_t(\theta), 1 - \epsilon, 1 + \epsilon)\hat{A}_t)] \tag{4}$$

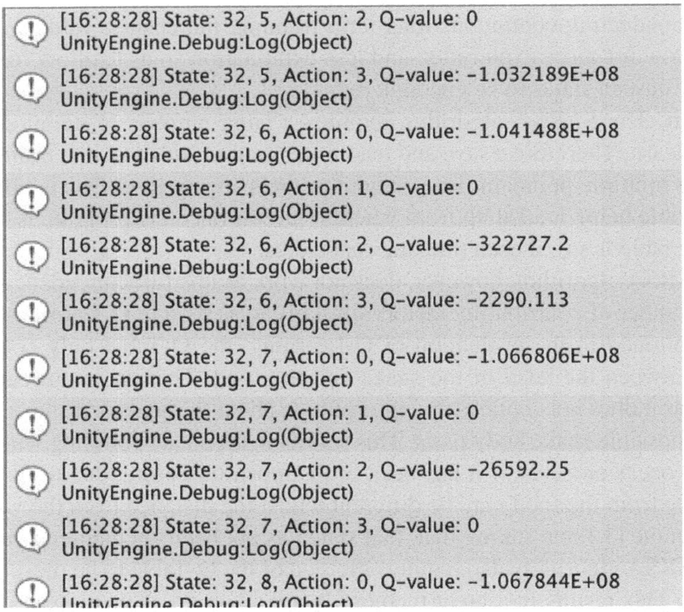

Fig. 11 A snippet of Q values within the Q-Table.

The above equation describes the objective of PPO. It makes use of Trust Region updates for compatibility with Stochastic Gradient Descent and does not require the need to create adaptive updates. θ is the policy parameter, $\hat{E}$ is the empirical expectation over time steps, r_t is the ratio probability under the new and old policies respectively, $\hat{A}_t$ is the estimated advantage at time t, and ϵ is a hyper-parameter, usually 0.1 or 0.2.

ML-Agents. ML-Agents [18] is an open-source machine learning toolkit for Unity that trains agents using reinforcement learning, imitation learning, neuro-evolution and other machine learning methods through a Python API. In our utilization of this toolkit, we made nine environments of the game, each of which could be played by the training agents asynchronously and in parallel, allowing for much faster training times. ML-Agent's implementation of PPO was applied to our game environment.

There were two possible designs for episode length. The short term design was with regard to Environment Reward, where the episode gets ended every time the agent either hits the border or collects food. When the agent dies, hitting the border, the game is finished with no reward. When the agent succeeds, collecting the food, the game is terminated with a reward of 1. On the other hand, the long term design is with respect to Cumulative Reward where the episode gets ended only when the agent hits a border, and whenever the agent collects a food a reward of 1 is accumulated. In summary, in the short term case, a boolean reward indicating either success or failure would be finalized. Meanwhile, in the long term case, the agent would end with a reward indicating the length of the agent's body. The short term design was

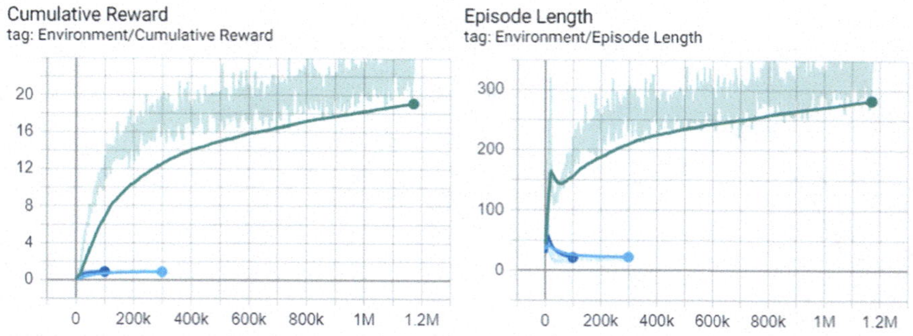

(a) Environment

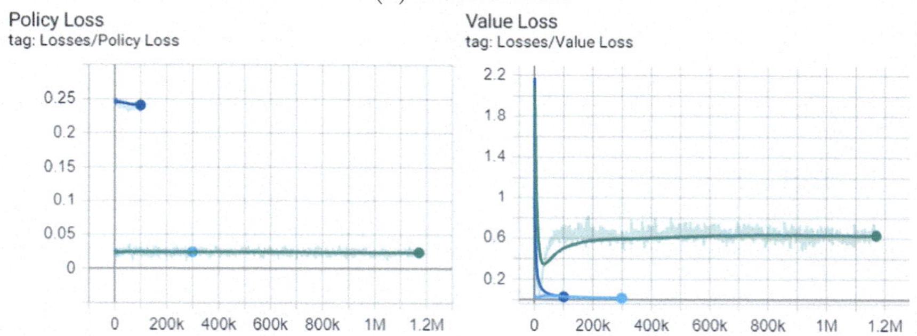

(b) Loss

Fig. 12 Long term episode (green) versus short term episodes (blues)

favoured as it allowed for faster training times as well as a more robust visualization of convergence in learning via graphs, by showing the average reward per step (Fig. 12).

Three types of observations were ensued. Firstly we implemented a vector-based approach, where position and direction vector observations of the environment were fed into the learning model. Secondly, we took a CNN-based approach, where the agents each had a camera overlooking their environment. Based on the frame they would perceive, they would determine the best possible action for their next step. Finally, we pursued a raycasting based approach, where rays were cast in the possible directions the agent could move, and the distance at which the rays intersect obstacles or food were observed. It is worth noting that the first approach solves the game with respect to collecting the food and avoiding borders only, whereas the CNN and raycasting approaches aim to extend to also cater for avoiding the tail.

Vector Approach. Two implementations of a vector-based approach were trained successfully. The first successful implementation (Without border observations) consisted of 6 observations for; the target food position, the agent's facing direction and the agent's current position. Note that each position has two observations, each with respect to the grid's x and y-axis. Since it only acknowledged these position/direction vectors, the agent struggled when it came too close to the borders. The second imple-

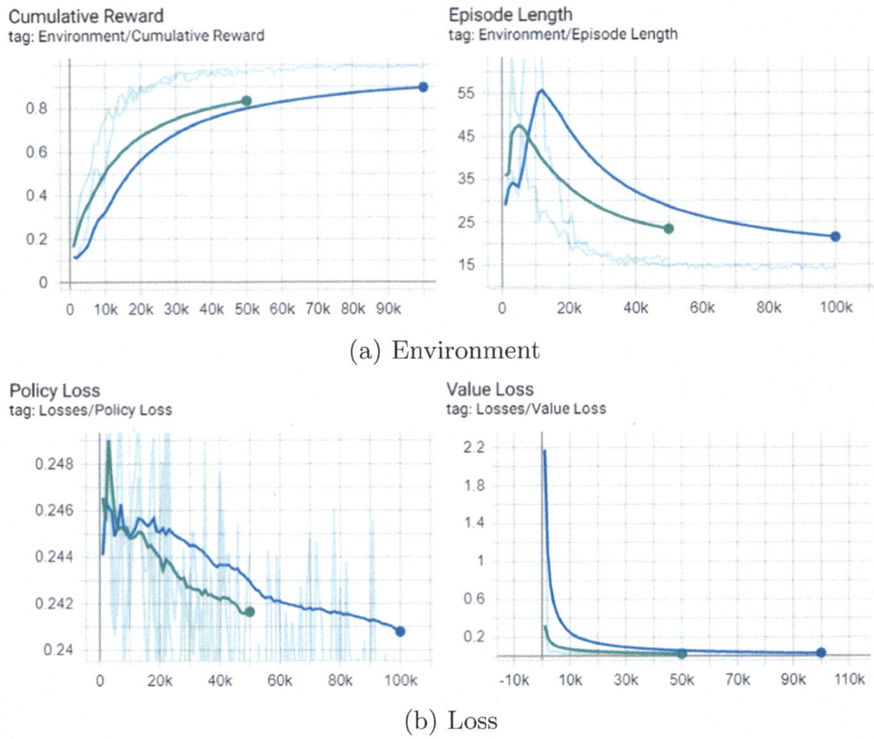

Fig. 13 Vector approach, without (green) versus with border observations (blue)

mentation (With border observations) improved on this via an additional four observations to set the bounds of the map; the top border y, the bottom border y, the right border x and the left border x (Fig. 13). With border observations, it started learning at a slower rate (due to a larger observation vector size) but converged at a higher reward. The episode length peaked later and higher, due to it firstly started with avoiding the border quite drastically, then tuning its avoidance to avoid it at more close quarters. The value and policy loss of With border observations also stabilize later than Without border observations. Note that the policy loss correlates to how much the policy is changing, which is expected to decrease in a successful training session. The value loss correlates to how well the model can predict the value of each state, which should increase while the agent is learning and decrease when stabilizing their reward. Now that we achieved an agent that was able to navigate towards the food while avoiding the borders, solving the problem of its tail is next.

CNN Approach. The problem was first approached through a Convolutional Neural Network (CNN) approach, using the ML-Agent's feature of attaching a visual sensor the agent. Each agent was connected virtually via a sensor to a camera overlooking their environment area. In the first trial, we aimed to speed up training time by taking the visual input at a low resolution of 96×54 and only considering one channel, in grayscale. After a long amount of time, its reward rate still couldn't converge (Fig. 14). The reward, episode length, policy and value loss were seen to spike too randomly and significantly for the agent to settle on consistent behaviour. At first, this was considered due to the input losing too much information at a low resolution and only considering one channel for grayscale. As a second experiment to improve on the first, the visual input was taken at a larger resolution of 256×144, and all three channels were considered, for colour, though it took much longer than the grayscale attempt due to its three channels (possibly making it three times as slow).

After running the program for two days, it did not achieve much and only covered half the amount of steps that the grayscale attempt achieved overnight (Fig. 14). Though it is worth noting that it did achieve a higher average reward and episode length, possibly due to more information. The policy loss peaked higher, and decreased at a larger gradient than the grayscale experiment, indicating the policy was changing. The value loss of the grayscale experiment peaked at around 200,000 steps, potentially indicating learning in the agents. It then decreased at 300,000, which would usually indicate the reward stabilizing, but as can be seen in the reward graph, everything was spiking far too much for stabilization. In fact, though the graphs seem smooth, note that the curves are smoothed averages of the real values, made for easier and more obvious comparison. It is the faded lines (the standard deviation) that show the real values throughout the experiments, which in this approach are all over the place.

In conclusion, the complexity of learning the environment through images was too high for the agents to learn. Though, if the experiment was left for longer, there could have been a potential for it to learn, but trying out different experiments in the CNN approach was too time costly. This lead us to approach this problem with less input complexity, with raycasting.

Raycasting Approach. Raycasting was implemented manually. The directions of rays were casted to the front of the agent, to the left of the agent, and to the right of the agent, because those were the three directions the agent was able to move at every step. For each direction, two rays originating from the agent were cast, totalling to six rays to cater for detecting obstacles and food. Layer masking was utilized so that in each direction, one ray was for detecting obstacles, and an other for detecting food. The ray length was set to the width of the game environment for optimal and realistically simulated perception.

For each ray, an observation was calculated conditionally using the distance from the agent to the ray intersection, as well as the ray length:

$$obs_{(dir,obj)} = \begin{cases} 1 - \frac{dist_{(dir,obj)}}{rayLength_{(dir,obj)}}, & \text{if objHit}_{(dir)} \\ 0, & \text{otherwise} \end{cases}$$

This general formula was applied to the forward, left and right rays for detecting obstacles and food respectively. As an example, the observation for when an obstacle is in front of the agent was calculated with:

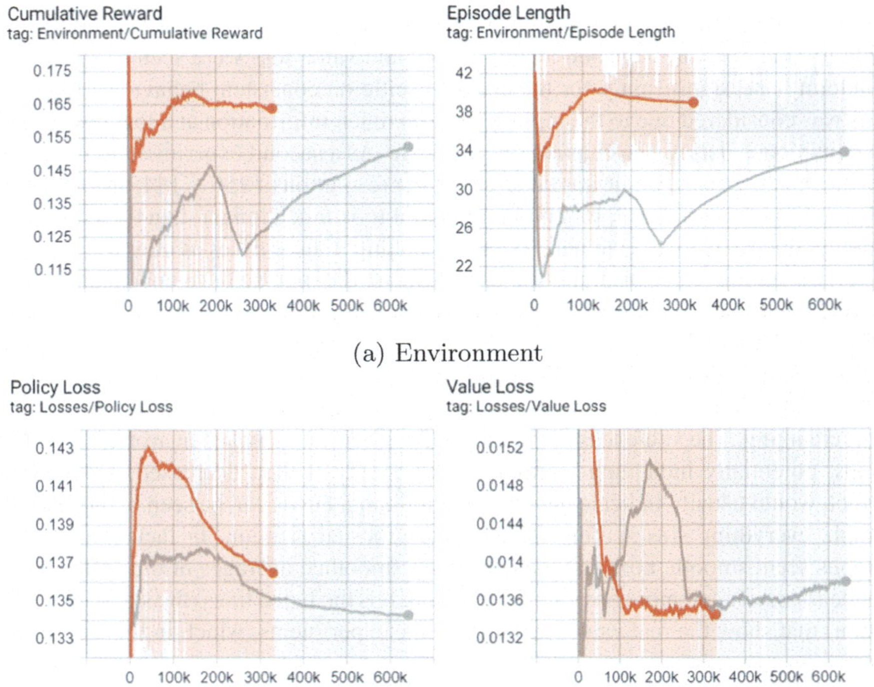

(b) Loss

Fig. 14 CNN approach, grayscale 96 × 54 (gray) versus color 256 × 144 (orange)

$$obs_{(forward,obs)} = \begin{cases} 1 - \frac{dist_{(forward,obs)}}{rayLength_{(forward,obs)}}, & \text{if objHit}_{(forward)} \\ 0, & \text{otherwise} \end{cases}$$

The observation is set to zero when no object is detected in front of the snake. The observation is set to some number ranging between 0–1 if a ray intersection does occur. The closer the value is to 1, the closer the distance of the obstacle in front of the agent, and vice versa. Note that the distance was divided by the ray length to normalize the observations, in attempts to make it easier for the neural network to adjust its weights. The fraction in the formula was subtracted from 1 to invert the observation in attempts to assist neural activation.

Raycasting successfully solved avoiding the tail. The first experiment (Without grid direction) was done strictly with respect to the six rays described above. Though its learning was decent, it was not that stable. During training, it was realised that the game object of the agent rotated on every turn and therefore so would the rays, making it so that the agent had no clear concept of what direction it was looking at with respect to the grid. Another investigation was carried out (With grid direction) to include two observations for the x and y of the agent's direction with respect to the grid. The reward rate With grid direction was seen to reach higher levels than Without grid direction, and the convergence had more stability (Fig. 15). The policy losses were seen to change quite frequently due to the observations continually changing with each turn of the agent. The value loss, on the other hand, dips at about 10,000 steps (exploring) then increases (learning) until 50,000 steps. It then starts to go down, indicating that the reward is stabilizing slowly.

The episode length of With direction was much shorter than Without grid direction, due to it having more control for movement. The experiment without grid direction seems to have adopted a less efficient random-seeming strategy for seeking out the food, in order to cater for its lack of knowledge in grid direction. Though With grid direction was notably better, it was seen that the agent would lose when it comes to body parts in front of it. It was theorized that this would happen because the agent didn't have enough time to practice with a long body length, and therefore does not experience the scenario of body parts in front of it as often as the scenario of body parts left or right of it. In attempts to improve on this, a third experiment was carried out with walls. Walls were placed on the upper area of the map and set with the same properties as the agent's body parts. As can be seen in Fig. 15, With walls converges at a lower reward than with walls (Known as With grid direction in the figure). This is due to the wall posing more of a challenging environment, and therefore causing more frequent deaths. The agent wasn't able to improve much, with its focus shifted to the static wall rather than food and its moving body parts.

3.4 Other AI

Other than reinforcement learning, there are a variety of different methods to approach developing an AI for the game of snake. An AI with heuristics, for example through raycasting and fuzzy logic, could be imperatively designed to tell the agent to avoid the obstacles when too close or move towards the food objects when in sight. It is worth noting that rule-based AI is ineffective when it comes to a frequent change in the environment. A better alternative is AI search algorithms because they are able to search for possible paths and can be executed everytime everytime an obstacles position is updated. As an example of a search algorithm for snake, we implemented A* pathfinding.

A* Pathfinding. A* pathfinding works by splitting an environment into units. Through these units, a variety of paths with respect to avoiding obstacles are computed to a target, and then through comparison, the paths are filtered down until the shortest path is found. The agent would then pursue through the estimated shortest path until the target is reached. In this algorithm, there are various heuristics for path searching such as Euclidean, Manhattan and Diagonal Manhattan distance.

We utilized Aron Granberg's A* Pathfinding Project for Unity [19]. It was made that at a frequency of 0.1 s, the agent would search for a new shortest path with respect to obstacles. The utilized pathfinding implementation did not support the discrete movement, and so we developed a function for movement with respect to

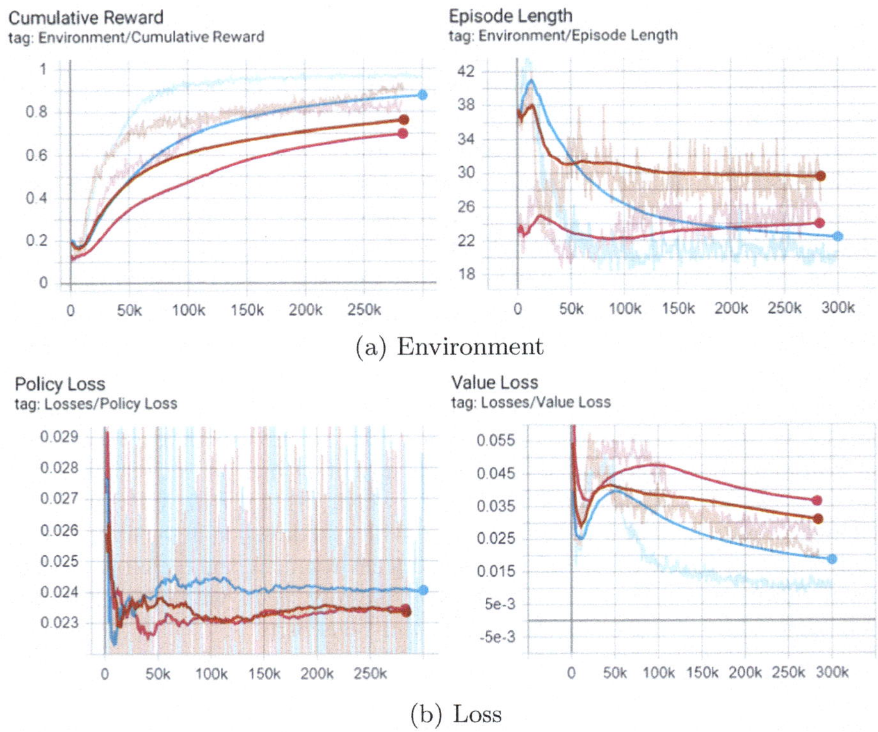

Fig. 15 Raycasting approach, without grid direction (orange) versus with grid direction (blue) versus with wall (pink)

the shortest path's waypoints. The agent was made to pursue through the shortest path by moving in the direction of the closest waypoint. After reaching that waypoint, it then moves in the direction of the next updated waypoint. The direction to closest waypoint was computed via the displacement from the agent's head location to the waypoint location. It was seen that though the agent is able to play the game, the agent doesn't traverse the shortest path as precisely as possible and hence sometimes overshoots the path in tight spaces when it focuses on a specific waypoint too far away. It doesn't play the game to its optimal capacity. In order to improve this, we could reimplement A* pathfinding by developing a version of the algorithm more specific to our discrete movement requirements.

4 Conclusion

We have implemented Snake as initially created in 1976. On top of which, different reinforcement learning algorithms were implemented, and their results were evaluated. The reinforcement learning algorithms Q-Learning and SARSA for variously discussed reasons did not produce optimal results and had various shortcomings. Differently, Proximal Policy Optimization (PPO) reinforcement learning algorithm outperformed the other implemented algorithms and gave near-optimal results in shorter periods of time. PPO was approached in three ways. The vector approach was able to play the game of snake, but only with respect to collecting food. The CNN approach did not produce optimal results due to input complexity. Finally, the raycasting approach simulated most intelligence as it was able to learn how to avoid obstacles and collect food. This may be improved with more rays and the introduction of sphere casting for better obstacle detection at a close proximity. In addition, A* Pathfinding was also implemented, which achieved strong performance but not better than the two PPO approaches.

References

1. Anunpattana, P., Panumate, C., Iida, H.: Finding comfortable settings of snake game using game refinement measurement. In: Advances in Computer Science and Ubiquitous Computing, pp. 66–73 (2016)
2. Jost, J.,, Li, W.: Reinforcement learning in complementarity game and population dynamics. Phys. Rev. E, Stat. Nonlinear Soft Matter Phys. **89**(2), 022113 (2014). ISSN: 15393755. http://search.proquest.com/docview/1639976188/
3. Kaelbling, L.P., Littman, M.L., Moore, A.W.: Reinforcement learning: a survey. In: CoRR cs.AI/9605103 (1996). https://arxiv.org/abs/cs/9605103
4. Sutton, R.S., Barto, A.G.: Reinforcement learning: an introduction. IEEE Trans. Neural Netw. **16**, 285–286 (1988)
5. Amiri, R., Mehrpouyan, H., Fridman, L., Mallik, R.K., Nallanathan, A., Matolak, D.: A machine learning approach for power allocation in HetNets considering QoS. In: 2018 IEEE International Conference on Communications (ICC), pp 1–7, May 2018. IEEE. https://ieeexplore.ieee.org/abstract/document/8422864
6. Zheng, Y.: Reinforcement learning and video games (2019). arXiv: 1909.04751 [cs.LG]
7. Silver, D., et al.: Mastering Chess and Shogi by self-play with a general reinforcement learning algorithm (2017). arXiv: 1712.01815 [cs.AI]
8. Mnih, V., et al.: Playing Atari with Deep Reinforcement Learning (2013). arXiv: 1312.5602 [cs.LG]
9. Ma, B., Tang, M.X., Zhang, J.: Exploration of reinforcement learning to SNAKE (2016)
10. Wei, Z., et al.: Autonomous agents in Snake Game via Deep Reinforcement Learning. In: 2018 IEEE International conference on Agents (ICA), pp 20–25, July 2018. https://doi.org/10.1109/AGENTS.2018.8460004.
11. Watkins, CJCH.: Learning from delayed rewards (1989)
12. Manju, S., Punithavalli, M.: An analysis of Q-learning algorithms with strategies of reward function. Int. J. Comput. Sci. Eng. 3 (2011)
13. Watkins, C.J.C.H., Dayan, P.: Q-learning. In: Machine learning **8**(3), 279–292 (1992). ISSN: 1573-0565. https://doi.org/10.1007/BF00992698
14. Xu, D., et al.: Path planning method combining depth learning and Sarsa Algorithm'. In: 2017 10th International Symposium on Computational Intelligence and Design (ISCID), vol. 2. pp. 77–82, Dec 2017. https://doi.org/10.1109/ISCID.2017.145
15. Chettibi, S., Chikhi, S.: An adaptive energy-aware routing protocol for MANETs using the SARSA reinforcement learning algorithm. In: 2012 IEEE Conference on Evolving and Adaptive Intelligent Systems. pp. 84-89, May 2012. https://doi.org/10.1109/EAIS.2012.6232810

16. Iima, H., Kuroe, Y.: Swarm reinforcement learning algorithms based on Sarsa method. In: 2008 SICE Annual Conference, Aug 2008, pp 2045–2049. https://doi.org/10.1109/SICE.2008.4654998
17. Schulman, J., et al.: Proximal policy optimization algorithms (2017). arXiv: 1707.06347 [cs.LG]
18. Juliani, A., et al.: Unity: a general platform for Intelligent Agents (2018). arXiv: 1809.02627 [cs.LG]
19. Aron Granberg's Pathfinding Project. https://arongranberg.com/astar/. Accessed: 2020-02-05

An Approach Towards Architecture-Independent Output for Generative Networks: Texturing Aerial Town Maps for Roleplaying Games

Gianfranco Siracusa, Dylan Seychell, and Mark Bugeja

Abstract Generative Adversarial Networks (GANs) have benefited of significant attention from the research community in recent years. The advancements in the field have led to high-quality results being achieved, but the computational requirements for matching these results have increased, making high-end methods unattainable by researchers with more limited equipment working on domains where pre-trained models for transfer learning are not available. A new system is presented for combining the output of small-sized, easily trained conditional generative adversarial networks to synthesise output at much higher resolution, even with small datasets, by applying a blending technique that uses redundant colour information to mend seams between network output that is spatially adjacent in the final image. The system was found to be capable of generating structurally consistent output, although it exhibits some defects where there are large areas of the same feature. The system creates output with average user opinion score of 3.5 on a scale from 1 to 5, which was generally judged to be an improvement over plain procedural output by users. The algorithm parameters allow the user to balance between performance, blending efficiency and sharpness of detail.

Keywords Generative adversarial networks · Game content generation · Seamless tiling

A different version of this work was published in 2021 IEEE Conference on Games [1].

G. Siracusa (✉) · D. Seychell · M. Bugeja
University of Malta, Msida, Malta
e-mail: gianfranco.siracusa.00@um.edu.mt
URL: https://www.um.edu.mt/ict

A. Dingli et al. (eds.), *Disruptive Technologies in Media, Arts and Design*, Lecture Notes in Networks and Systems 382, https://doi.org/10.1007/978-3-030-93780-5_13

1 Introduction

Aerial town maps are an essential part of any roleplaying game. Besides serving as a gameplay component for finding one's way to objectives and important landmarks, players who enjoy immersion in the game world look towards maps to provide context and character for the town or area they find themselves in.

Town maps are sought after by various creative categories, including books, campaign settings, quests and adventures, stand-alone art, and digital games. Several tools and software (see Sect. 2.1) to assist with the drawing, or to procedurally generate, town (and land) maps have been built to address the need that fantasy authors have in creating fantasy maps of their own.

In light of the recent advancements on AI-based image generation methods in other domains, an attempt is made to apply this generative approach to overcome the limitations of modern software in this regard: the simplistic texturing of procedural design and the limitation on variety imposed by sprite-based rendering.

Generative Adversarial Networks (GANs) were originally proposed by Goodfellow et al. in [2] and have been researched heavily since. Improvements include the development of several architectural variants such as deep convolutional GANs [3] for high-resolution image synthesis, conditional GANs [4] for image-to-image translation, and Information Maximising GANs [5] for feature disentanglement, to mention a few.

Research has been carried out on other improvements including normalisation techniques [6], objective functions [7], generator and discriminator balancing [8], unrolling discriminator gradients to anticipate changes and minimise mode collapse [9], and many more.

Over time, these generative approaches have been applied to image generation [10], sound effects synthesis [11], paired [12] and unpaired [13] image-to-image translation, synthesis of fake data towards data augmentation [14], super-resolution [15], stitching panorama images [16], image extension [17], image colourisation [18] and game compression [19], amongst others.

The challenge with current GAN methods is that the resolution of the output is heavily dependent on the architecture. The restriction originates from the fully connected layers that latent input vectors are usually linked to the network through. This presents a problem, given that higher resolutions demand more processing power and a longer duration for training.

For example, Karras et al. [20] report leaving their model run on 8 NVIDIA Tesla V100 GPUs for one whole week to achieve the outstanding results demonstrated in their paper at a resolution of 1024×1024.[1] To the authors' best knowledge, output resolutions with more than around 1×10^6 pixels are extremely rare. The few that exceed this resolution generally employ few-shot training or pyramidal refinement.

[1] A live demonstration of Karras' method at work on generating realistic images of faces can be found at *This person does not exist: Faces dreamed up by a GAN*. https://thispersondoesnotexist.com/.

This imposes a practical limit on the resolutions that can be achieved from generative adversarial networks. Not only is a 1k resolution image not up to today's display unit and file sharing standards, but even achieving this resolution is not a feasible task in many circumstances. Most researchers without access to high-end resources therefore have to settle for even lower resolutions to get decent results in a reasonable time period.

It would therefore be helpful to find a method by which one could achieve higher resolutions without suffering from the long training times and computational power required from such architectures. This paper introduces a method using which high resolution images can be patched together from tiles synthesised with networks that have a significantly smaller intrinsic output resolution.

The proposed system performs image-to-image translation from colour class labels to fully textured images, where both input and output have arbitrary – and not necessarily identical–size. In order to demonstrate this, however, a context is needed, preferably from a continuous domain (such as terrain, maps, natural scenes, etc.) as opposed to a well-defined objects (such as faces, cats, bedrooms, etc.).

The drawing of artistic maps was eventually chosen as the context for investigating effectiveness of the system that will be described here. The success of the proposed method will be in part evaluated by the artistic appeal of aerial town maps it produces, as well as a number of metrics appropriate for the task.

2 Related Work

2.1 Current Mapping Options

Traditional approaches to fantasy town mapping involve procedural design.[2,3,4] The results achieved by such methods are of remarkable quality in terms of layout; however, procedural methods are generally limited in producing texturing detail, resulting in plain colours or at best simple textures.

Other approaches use a fixed library of sprites prepared by a human artist to represent features like buildings and vegetation, placing these at locations determined manually [5] or by a procedural design step.[6] This overcomes the issue with texturing detail, but in this case the variety is limited by the number of sprites available, and the believability of the output tends to suffer if the same sprite is used too often.

[2] Watabou's *Medieval Fantasy City Generator*. https://watabou.itch.io/medieval-fantasy-city-generator.

[3] Watabou's *Village Generator*. https://watabou.itch.io/village-generator.

[4] Drachenzahn's *Roleplaying City Map Generator* (Original link no longer maintained).

[5] *Inkarnate*. https://inkarnate.com/.

[6] *Cityographer*. https://cityographer.com/.

2.2 Generative Adversarial Networks

Generative Adversarial Networks (GANs) consist of two components, a Generator *f(G)* and a Discriminator *f(D)*, which are in competition with each other. The Generator's aim is to learn a target distribution and hence reproduce an output that is so realistic that the Discriminator cannot distinguish it from a real sample. On the other hand, the Discriminator's aim is to learn how to tell which of its inputs were taken from the real dataset and which were created by the Generator.

Both components of this combined model provide feedback to the each other for their improvement. A temporary success of the Generator in fooling the the Discriminator forces the latter to counter by recognising the pattern in the output that is making it look realistic, and the Generator responds by exploring other patterns that make its output look real, always improving its ability to mimic the real distribution.

Because of this adversarial nature, training GANs is considered as a zero-sum scenario that follows the core principles of a Nash Equilibrium in game theory [21]. Neither of the two components should be allowed to dominate over the other as this would compromise the success of the entire system. Hence, the architectures of the two models have to be designed carefully to prevent one component from being so powerful as to not give chance for the other component to catch up and restore the cyclic improvement process.

2.3 Conditional Generative Adversarial Networks

Out of all the variants of Generative Adversarial Networks that have been developed over time, conditional GANs are the most closely related to the work being presented here. Conditional Generative Adversarial Networks, or cGANs, are generative models that are capable of producing multiple classes of output in the same image based on some conditioning information that is provided as part of or in addition to the main input data.

Conditional GANs were introduced by Mirza et al. in [4], where their potential was demonstrated in the synthesis of specific handwritten digits and in image-to-text generation. In the former, conditioning occurred by a single value which determined what digit the network will generate next, alongside the random vector that determined the actual appearance of the digit.

Later Isola et al., [12] expanded the same concept by passing an entire image as a conditioning input, instead of a single value. This image acts as a segmentation mask that tells the network which object or feature occupies which space in the image, allowing the network to generate an output image in a specific layout determined by the user.

The authors used this technique to convert images from one domain to another (a task known as *image-to-image translation*), such as constructing an image of a facade from a segmentation mask provided in plain colours, colourising greyscale images, or converting between day-time and night-time representations of the same scenery, amongst others. The technique was widely adopted by the public, who used it for numerous projects of their own [12].

A further addition towards the image-to-image translation task was provided by [13], who introduced cycle-consistent conditional GANs for datasets where the same image is not available in both domains that one wishes to exchange between. This task, known as *unpaired* image-to-image translation, frequently gives inferior results than the paired translation task described above, but is nevertheless an important research area given that obtaining the same image in both source and target domains may be unfeasible or time-consuming in practice (for example, taking a photograph of the same scenery in different seasons using identical viewpoints).

To achieve this aim, the authors proposed a technique whereby separate generators and discriminators exist for each translation direction (from source to destination and vice-versa) and the losses of the generated output—even though not directly correlated— can be estimated from similar points in feature space. This technique was successful in converting between summer and winter scenes, or between zebras and horses. The authors also demonstrated that the concept can be extended to individual frames of a clip.

The following are some recent advancements in the field of image synthesis from semantic segmentation maps.

Improvements to Image-to-Image Translation [22]. Based on previous work related to image-to-image translation [12], the authors of this paper introduced several improvements to support higher-resolution output as well as stabilize adversarial training. Image synthesis is taken in separate steps using different generators in a pyramidal structure, each taking the previous output and refining it to the next resolution factor. Multiple discriminators operating at different scales are employed in order to assess features of varying sizes and also to minimise the occurrence of mode collapse. Feature matching loss from intermediate discriminator layers serves to extract better gradients to follow.

Natural Scene Synthesis by Semantic Manipulation [6]. This work investigates methods for creating realistic natural scenes by drawing a semantic map of the features constituting the scene. An interactive demonstration of the method is present online.[7] The main contribution of this work is the definition of a new normalisation technique known as spatial normalisation. This algorithm addresses the issue that is commonly encountered in image-to-image translation, which occurs when a single patch in the discriminator consists entirely of the same label. All patches that have this characteristic normalise to the same values with traditional batch normalisation, which would lead to loss of semantic information that is essential for consistent and accurate colouring.

Natural Scene Manipulation by Transient Attributes [23]. In this other work related to image generation from semantic information, which is somewhat more oriented towards unpaired translation and implements multimodal operation, nature

[7] *GauGAN*. http://nvidia-research-mingyuliu.com/gaugan/.

scenes are extracted from two datasets: ADE20K [8] and Transient Attributes.[9] The images are given annotations across 150 semantic classes and 40 transient attributes such as season, light conditions, time, weather and so on. While the classes allow the layout of the image to be defined by the user, the transient attributes enable control over various properties related to the style of the output.

2.4 Spatial Generative Adversarial Networks

The work presented in this paper is not the first to try to approach the problem of network architectures rigidly imposing output resolution. A variant of generative networks that are entirely convolutional has been devised to extract patterns learnt from a single image or small set of images and then generate an image of arbitrary size that exhibits very similar high-level structure and low-level detail of the source [24]. One use case of Spatial GANs towards the generation of terrain is described in the next section.

Although the results obtained by this work are fit for the intended purpose, the technique is specifically designed with textures in mind and therefore produces patterns that are highly repetitive. Consequently, it is unsuited for domains where the output is expected to be more spatially structured, leaving the need for other models to address the issue of arbitrary output in such use cases.

2.5 Generation of Maps with Machine Learning

Terrain, Vegetation and Cities with CNN and GAN [25]. This work takes a single street-level image of a real city and attempts to generate a 3D model including terrain, buildings, streets and vegetation. The aim of the authors is to produce a 3D model that has similar appearance and characteristics to the reference city, not to generate an accurate digital representation of the exact city. The authors use a Generative Adversarial Network to synthesise a heightmap and corresponding terrain from a latent vector extracted from a segmentation map of the street-level image. A separate Convolutional Neural Network extracts properties and characteristics related to the city from the image, such as street materials, building density, facade types, vegetation types, and so on. These characteristics are used to create, texture and place 3d models on the terrain generated with the GAN.

Vegetation with CNN [26]. This work aims to use a Convolutional Neural Network to learn a vegetation distribution from an exemplar terrain and use it to quickly populate a target terrain with the same pattern of vegetation. The benefit of this is to speed up the time-consuming task that 3d artists encounter when placing vegetation in scenes.

[8] *ADE20K* dataset. https://groups.csail.mit.edu/vision/datasets/ADE20K/.

[9] *Transient Attributes* dataset. http://transattr.cs.brown.edu/.

The system developed by the authors allows artists to place vegetation on a simple terrain (such as plain elevations), letting the CNN learn what kind of plants grow at which altitudes and orientation (with respect to light). The CNN then quickly places vegetation on more complex terrain, consisting of more frequent altitude variations, which would take much longer to place by hand.

Terrain [27]. The authors of this paper utilise Spatial GANs (see Sect. 2.4 above) to learn patterns from Shuttle Radar Topography Mission (SRTM) elevation data and use the trained model to create heightmaps that can be used in 3d scenes and games. Spatial GANs are ideal for this purpose because they can learn structured patterns that repeat over different levels of scale, as elevation does. In fact, the authors report that their technique leads to superior heightmaps than those which are derived from noise-based and fractal-based procedural methods, which are the traditional techniques used for the purpose. A limitation of these is that they tend to become repetitive and lack the structured diversity of real terrain. Another point in favour of Spatial GANs—as opposed to other machine learning methods—for continuous domains such as terrain is that their resolution-unbound nature is not detrimental to the semantics of the output.

2.6 Blending Techniques

Other works have employed some extent of stitching or blending between separate images. Nevertheless, to the best knowledge of the authors, the approach and intention of the methodology presented in this paper differs from previously published work.

Blending two sources [28]. GANs have been applied to blend parts of two images together in a more natural way than can be achieved with traditional Gaussian-Poisson-based techniques. The authors achieve this by constructing two models to address separate constraints: one for colour and one for gradient. The colour constraint is determined by means of a low-resolution network, which approximates the blended colours of the two images while avoiding issues related with training more complex networks. The gradient constraint serves to restore sharp edges and contrast missing from the low-resolution output. The authors use a Laplacian pyramid of generators to upscale the low-resolution predicted blend, while algorithmically exploring the search space to enforce the system to pick solutions that satisfy the gradient constraints.

Stitching photographs into a panorama view [16]. GANs were also utilised to stitch together a panorama view from a set (front, back, left, right) of contiguous or nearly contiguous images of the same scene, even if they have different fields of view, such as if they were shot with different cameras or at varying zoom levels. The method proposed in this paper first extracts features from the images and feeds them to a convolutional neural network, which performs regression on the features and returns an estimate of the FoV of each. This result is then used to put all images to scale if necessary. The generative component consists of a pyramid of three generators, each taking the output of the previous generator and upscaling it by a factor of two. Each generator is assessed by its own discriminator and those of previous levels to provide multilevel feedback on the final image.

3 Methodology

3.1 Dataset

Source maps. A small amount of high-resolution medieval fantasy town maps drawn by human artists were selected in such a way as to provide a variety of artistic styles and features. Any labels were edited out of the images, given that GANs may learn text-like symbols but currently are not capable of constructing meaningful text. The edited images were cropped to regular aspect ratios (1:1 or 3:2).

Features. The selected maps were segmented by hand into 34 predetermined features, including various types of terrain, water, grass, vegetation, and buildings. It is rather optimistic to expect the network to reliably represent all of these features, especially those which occur rarely in the dataset. Nevertheless, all foreseen labels were included in the feature set as it is easier to selectively drop some features procedurally than re-label the dataset to add new ones.

Preprocessing. Before being run through the model, the full maps were cut into superimposed tiles at five different scale levels. The purpose of this scaling, as opposed to jittering (outlined in Sect. 3.2), is to provide model robustness against large changes in the size of map elements, given that the final model is expected to be able to represent settlements at different scales, from hamlets to towns to cities. Robustness to scaling also helps with the tiling algorithm described below (see Sect. 3.3), as this involves rendering the image at increasing scales. Combining these techniques, preprocessing produced a dataset of 2000 tiles from the full-size maps, which was split in a 75:25 ratio between training and test sets (Fig. 1).

3.2 Network Model

Data augmentation. The particular scenario that is being dealt with allows for several forms of dataset augmentation, given that aerial maps make sense even after being manipulated in any of the following manners: translation, rotation, scaling, and flip operations. Thus, before being submitted for training, tiles in the dataset are randomly rotated at 45 degrees intervals and flipped in either direction. In addition, the tiles were jittered to provide model robustness against small changes in scale, such as differences in building sizes. The hue of the real images (not the masks) was adjusted by a random value between $\pm 5\%$.

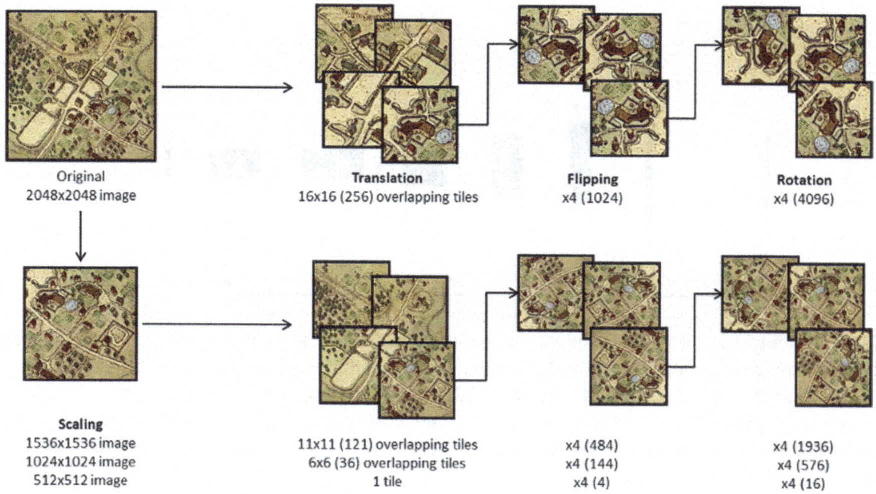

Fig. 1 Overview of methods used to produce tiles for the dataset. Tiling and data augmentation techniques allow comprehensive datasets to be created from a small number of hand-segmented maps. Featured map: Triboar village from Storm King's Thunder (Wizards of the Coast)

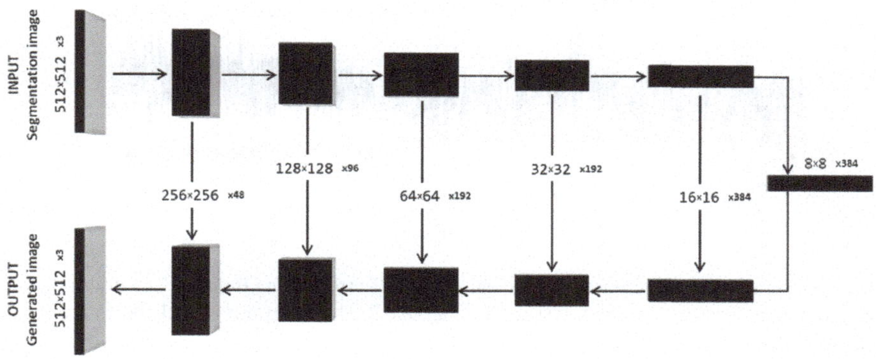

Fig. 2 Architecture of the generator used in this work

Foundation. The architecture for this research was based off the Tensorflow implementation [10] of the Pix2Pix algorithm developed by Isola et al. in their image-to-image translation work [12]. The original implementation uses 4×4 px kernels but these were changed to 5×5 px kernels as the closest symmetric kernel (Fig. 2).

Generator. The generator is a 6-layer downsampling and upsampling U-net with skip connections. All kernels have a size of 5×5 px and a stride of 2. The detailed architecture is depicted in Fig. 2.

[10] Tensorflow *Pix2Pix* Implementation. https://www.tensorflow.org/tutorials/generative/pix2pix/.

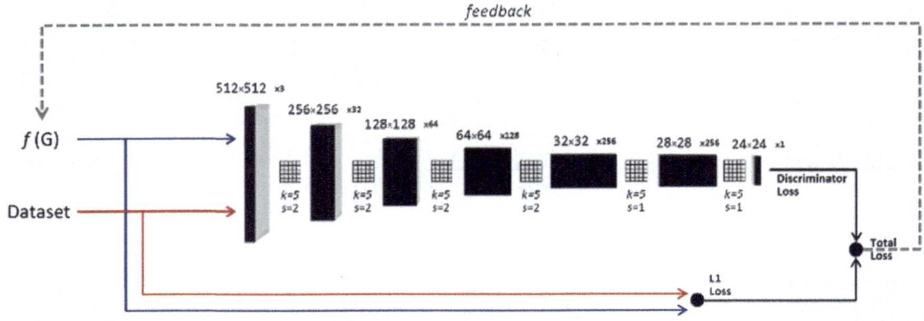

Fig. 3 Architecture of the discriminator used in this work. As described in the original paper for image-to-image translation [12], the discriminator takes in the output of the generator f(G) and real images from the dataset, attempting to distinguish between the two. The success of the discriminator, as well as the Mean Absolute Error between real images and generated versions (L1 error), is given as feedback for improvement of the generator

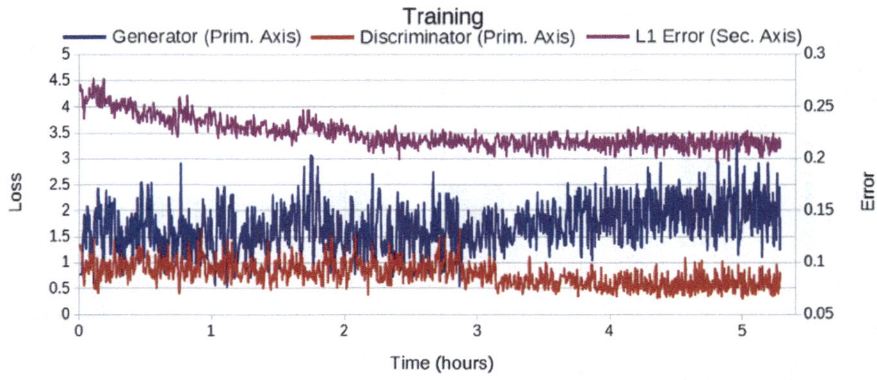

Fig. 4 Generator loss, Discriminator loss and system L1 error during training. The discriminator loss settles at around 0.7; generator loss settles at around 2. The L1 error is roughly 0.215 at 5 h and improves only slowly thereafter

Discriminators. The discriminator has 4 downsampling layers with a stride of 2, followed by 2 layers with a stride of 1. In either case, the kernel size is 5×5 px. The detailed architecture is depicted in Fig. 3. Each neuron in this discriminator has a receptive field of 189×189 px. This is meant to ensure structural consistency in overlapping windows of around one-third the size of the actual output.

Training. The network was trained on a batch size of 10 images, using Adam optimizers with a learning rate of 2×10^{-4} with $\beta_1 = 0.6$ for both the generator and discriminator. The learning rate was reduced by a factor of 2 after every 50 epochs. The model was trained for 200 epochs (just above 5 h). The training progress is shown in Fig. 4 and a sample of hidden layer activations is shown in Fig. 5.

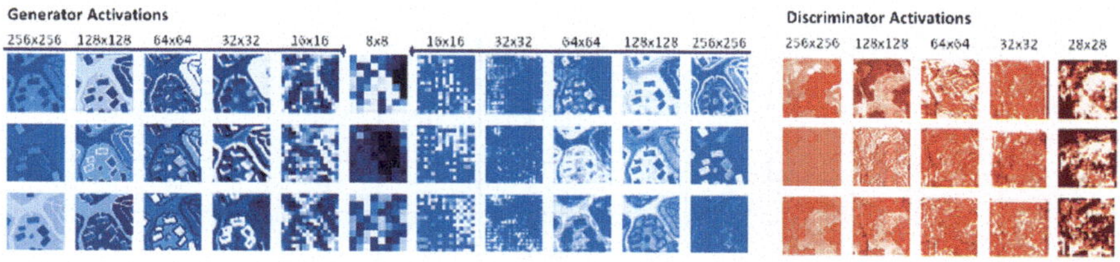

Fig. 5 Hidden layer activations

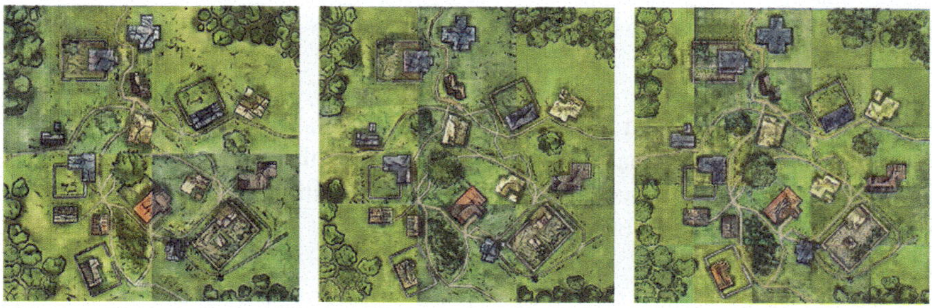

Fig. 6 Demonstration of seams when reconstituting the original image from tiles. The same image was passed through the network as a grid of 2 × 2 (left), 3 × 3 (centre) and 5 × 5 (right) tiles

3.3 Blending Algorithm

Problem Description. In order to produce an image of arbitrary size from the model described above, which has a fixed resolution of 512 × 512 px, the input has to be fed to the network as separate pieces of the specified dimension and subsequently reconstituted. However, since the network is processing the tiles separately, it has no information about which tiles are adjacent to the one currently being textured, and therefore cannot predict the colour of the pixels at the edges accurately enough to guarantee that these are consistent when the separate tiles are patched back together (Fig. 6).

Multiresolution Tiling. In order to stitch or blend different images together, there must be some extent of redundant information. In stitching, this is used to anchor the separate images in relation to each other. In this case, there is no issue about the position of the images as all tiles are contiguous and their position is known. Instead, redundant information about the colour of pixels at the edges is needed so that any errors are smoothed out. To this end, the image is run through the network more than once, each time at different multiples of the network resolution, so that the edges occur in different parts of the image when it is scaled back down to the original resolution (Fig. 7).

Pixel Fitness. Processing the input image at resolutions which are prime multiples of the architecture resolution ensures that the seams do not coincide. Hence, there is always at least one layer which has accurate colour information that can be used to

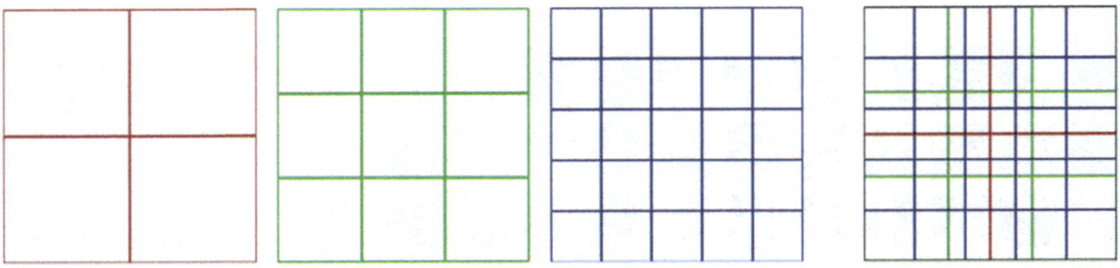

Fig. 7 High-level concept of the blending algorithm. Selecting prime numbers (2, 3 and 5) for the factor at which the image is subdivided before being passed through the network ensures that the seams do not occur at the same location in the image (far right)

determine the colour of that pixel in the final result. The contribution that the data in each layer has on the final pixel colour depends on the inverse of the Euclidian distance of that pixel, as we formulate in Eq. 2, from the centre of the tile in that particular layer, which we formulate in Eq. 1.

$$\chi_{(x,y,z)} = (\lfloor \frac{P_x}{T_W} \rfloor + 0.5) \times T_W \quad , \quad (\lfloor \frac{P_y}{T_H} \rfloor + 0.5) \times T_H \tag{1}$$

where P_x is the pixel's x-coordinate, P_y is the pixel's y-coordinate, T_W is the tile width, T_H is the tile height, x is the width dimension, y the height dimension and z is the channel dimension.

$$\omega_{(x,y,z)} = \sqrt{(P_x - \chi_x)^2 + (P_y - \chi_y)^2} \tag{2}$$

where χ_x and χ_y are the coordinates for the centre of the respective tile as determined with Eq. 1.

Layer Blending. Once the fitness values of a particular pixel in all layers is calculated, the lowest value is dropped (this is equivalent to discarding the contribution of a layer that is closest to an edge) and the remaining values are normalised so that they sum up to 1. The final colour of that pixel can then be determined by the sum of weighted RGB channel values, as we formulate in Eq. 3.

$$B_{(x,y)} = \sum_{z=0}^{n} \frac{\omega_{x,y,z}}{\sum_{z=0}^{n} \omega_{x,y,z}} \times RGB_{x,y,z} \tag{3}$$

where n is the number of layers, $RGB_{x,y,z}$ is the colour of a pixel at a specific position in a particular layer z and $\omega_{x,y,z}$ is the contribution of the layer pixel towards the colour of the same pixel in the final image.

3.4 Computational Resources

All computation required for this research was carried out on a privately owned Windows device possessing a 4.2GHz octacore processor, 32GB of RAM and an NVIDIA GTX1070 graphics card with 6GB of on-board graphical memory. Comparisons of the algorithm's performance under different parameters are given relative to a base case specified in the text instead of actual seconds.

4 Results

4.1 Sample Output

A set of segmented layouts for town maps was prepared and run through the system described above in order to judge the performance both visually as well with analytical methods described later in this section. The selected maps were derived from various sources, including a number of manually designed, procedurally generated and real-world layouts, listed below:

1. sampled map: Pinepass by David Barrentine,
2. a segmented full-sized map from the dataset,
3. a segmentation map which was manually designed using the tool developed for this research,[11]
4. large-scale terrain generated with procedural mode of the same,
5. terrain generated with a DCGAN trained on segmented maps,
6. output from Watabou's Medieval Fantasy City Generator,
7. output from Watabou's Village Generator,
8. output from Drachenzahn's Roleplaying City Map Generator, and
9. map of Dingli, Malta obtained from Google Maps.

Where necessary, the sources were adapted to match the colour palette that the network was trained on (Fig. 8).

4.2 Image Quality Metrics

While the assessment of image quality is notably a difficult task [29], usually left to methods involving human judgement such as user opinion scores, there are some factors that are considered typical of good quality in both photography and images.

The following image quality metrics were used in this analysis:

- **Mean Intensity** ($\bar{I}$)
- **Dynamic Range** (DR)
- **Histogram Flatness Measure** (HFM) [30]

[11] *Gantographer*, https://gstw20.eu.pythonanywhere.com/.

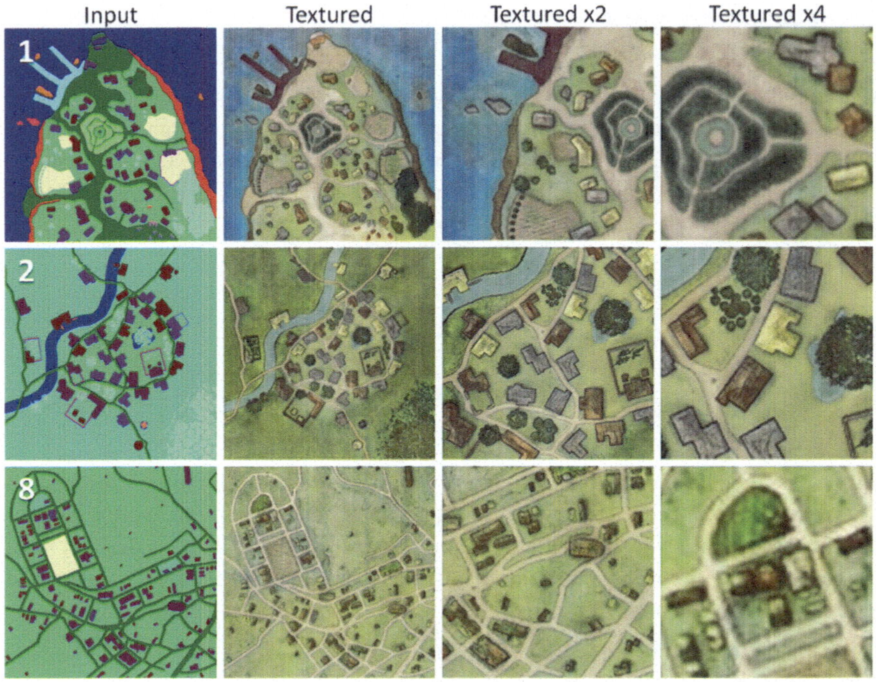

Fig. 8 A set of system output from manual or natural sources. The numbers correspond to the categories listed in Sect. 4.1. Examples of system output on the remaining categories—all from procedural sources—can be found in Fig. 17 in the Appendix

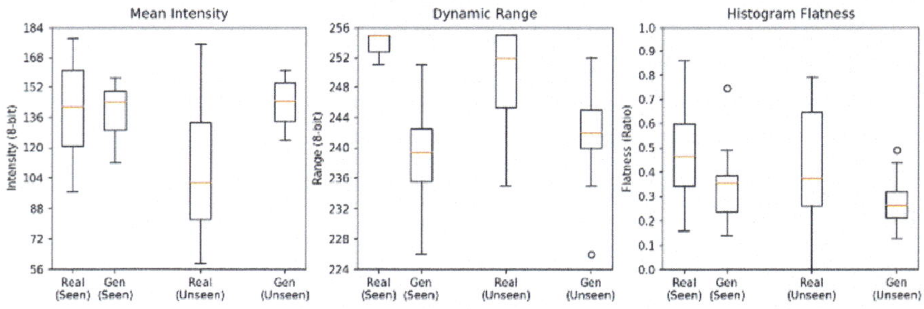

Fig. 9 Mean Intensity, Dynamic Range and Histogram Flatness Measure of various categories of input

The results of this analysis are shown in Fig. 9. The effect of changing system parameters on the same metrics is illustrated in Fig. 13.

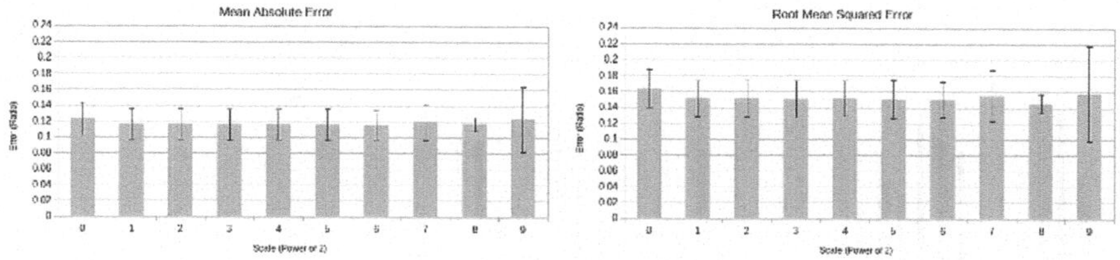

Fig. 10 MAE and MSE values for maps generated from the full-size dataset masks

4.3 *Reference Analysis*

The quality metrics mentioned above are general tests that assess appealing colour distributions rather than accuracy of the content. For assessing the content of an image, one needs to have the ground truth available. These methods of assessment are known as reference metrics. Although these methods are more specific and powerful, they restrict the analysis to the set of manually segmented artist maps only.

The following reference metrics were used in the analysis:

- Multilevel **Mean Absolute Error** (MAE) [30] and **Root Mean Squared Error** (RMSE)
- **Peak Signal to Noise Ratio** (PSNR) [30][12]
- **Structural Similarity** (SSIM) [33][13] and **Multiscale Structural Similarity** (MS-SSIM) [35]

The multilevel MAE and MSE values obtained on reference images at default parameters are shown in Fig. 10. The effect of changing system parameters on the PSNR, SSIM and MS-SSIM of the entire set of reference images is illustrated in Fig. 14.

4.4 *Segmentation Scores*

One common method of evaluation of conditional generative adversarial networks is through segmentation scores. Common segmentation metrics are mAP (mean Average Precision) and FID (Frèchet Inception Distance) used in [6], Accuracy and IoU (Intersection-over-Union *aka* Jaccard Index) used in [12], Inception Score (IS) and FID used in [23], among others.

The type of networks that are generally used for segmentation to derive the scores above are Fully Connected Networks (FCNs) [12], Cascaded Refinement Networks (CRNs) [22] and U-nets [22].

[12] PSNR is used as a metric in [16–18, 31, 32], among others. This work uses the Tensorflow implementation of PSNR.

[13] SSIM is used as a metric in [16, 31, 32, 34], among others. This work uses the Tensorflow implementation of SSIM and MS-SSIM.

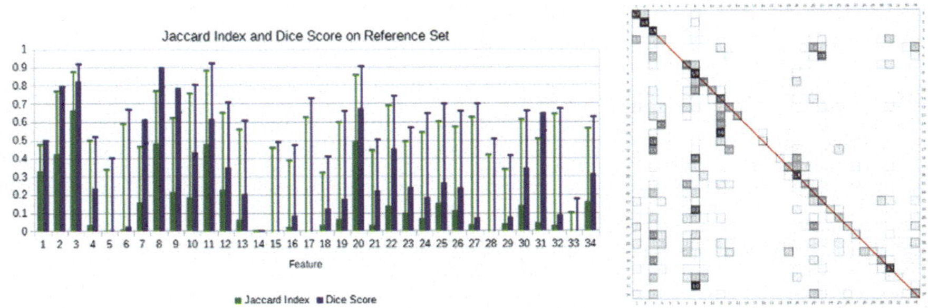

Fig. 11 Segmentation analysis on the reference set. **Left:** Jaccard Index and Dice Score for each class. The confidence of the segmentation model itself on the test set with the same metrics is shown as an ascending stem. **Right:** Confusion matrix between classes

A custom U-net segmentation model was trained to recognise map objects and textures in the dataset. The detailed architectures and methodology of the segmentation model is beyond the subject of this paper and will not be discussed here. However, it should be noted that this being a context for which no pre-trained models nor extensive datasets exist, it cannot be expected that the segmentation itself provides an optimal standard to compare results with.

Nevertheless, the scores obtained on generated images can still be used to assess the performance and characteristics of the texturing model described in this paper when interpreted in relation to the performance of the segmentation model on the ground truth images. The segmentation metrics chosen for this work are Intersection-over-Union (Jaccard Index) and F1 Score (Dice Score).

The generated images were separated in two different categories: reference (images generated from the segmentation maps that were manually labelled from artist maps selected for the dataset) and blind (images generated from other sources of segmentation maps, discussed in Sect. 4.1). This separation was necessary because it is expected that the texturing network performs somewhat better on the reference category, given that it has encountered these maps before, although as tiles not as a whole image.

Segmentation scores for the reference set are shown in Fig. 11, while scores for the blind set are shown in Fig. 12. The scores are listed by the capability of the segmentation model itself to recognize the various categories in Table 1.

4.5 User Opinion Scores

A survey was carried out in which respondents were asked to evaluate the visual appeal of maps generated by the proposed method. A total of 45 responses were collected. The majority of the audience consisted of individuals who had moderate experience in roleplaying games (25); a few were unfamiliar with the genre (5) while a good number were experienced players and content creators (15).

Table 1 Segmentation scores for reference and blind sets

Feature	Model accuracy (%)	Reference set	Blind set
House (Wood)	90.5	0.547	0.379
Path	89.8	0.745	0.459
Tree	88.2	0.583	0.484
Water (Shallow)	78.2	0.609	0.522
House (Thatch)	78.2	0.310	0.192
Grass (Light)	78.1	0.690	0.672
Remaining features	52.2 (0.3–71.8%)	0.15 (0.00–0.50)	0.06 (0.00–0.48)

Only the features with the highest reliability on the segmentation model are shown. The scores in this table represent the average of the Jaccard Index and the Dice Score for the respective feature and the respective set

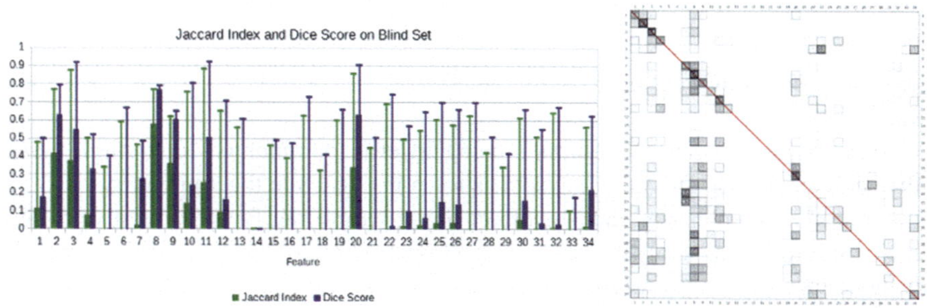

Fig. 12 Segmentation analysis on the blind set. **Left:** Jaccard Index and Dice Score for each class. The confidence of the segmentation model itself on the test set with the same metrics is shown as an ascending stem. **Right:** Confusion matrix between classes

The survey revealed that the majority of respondents (78%) gave a score of 3 or 4 out of a maximum of 5 for maps generated by the proposed method (mean user opinion score 3.53). The respondents only recognised maps textured by AI in 27% of the cases among a small but representative sources of maps, confusing them with hand-drawn maps in 53% of the cases and with paid fantasy map-making software in the remaining 20%.

5 Experiments

5.1 Algorithm Parameters

The main two parameters of the algorithm described in the Methodology are the number of layers rendered and the base tiling factor. For each layer, the number of tiles it is subdivided into is increased. For example, with 4 layers and a base tiling

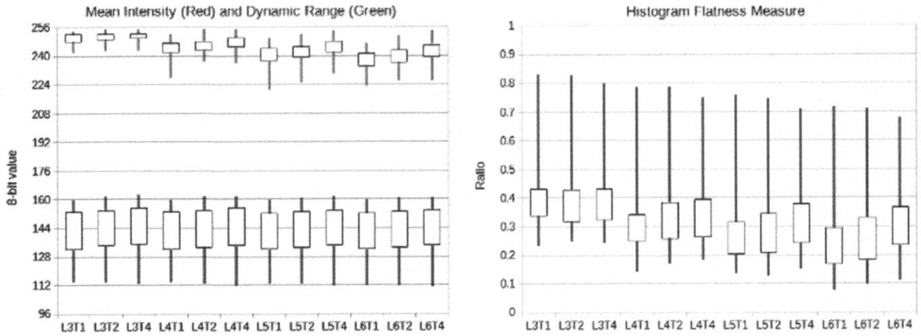

Fig. 13 $\bar{\text{I}}$, DR and HFM values for images processed with different settings. *L* stands for number of layers; *T* stands for tiling factor. Whiskers represent minimum and maximum, box represents first and third quartiles for each category

factor of 2, the individual layers will consist of 2×2, 3×3, 5×5 and 7×7 tiles respectively.

An example of the network output—using different parameters—on a single manually segmented map from the dataset is shown in Fig. 18 in the Appendix.

It should be emphasised that the effect of the base tiling factor is relative to the target resolution. The analysis was run on a target output of 1024×1024 px, which is equivalent to a tiling resolution of 2×2 for the architecture described above.

Figure 13 indicates a strong positive correlation (0.936 for $\bar{\text{I}}$, 0.954 for DR and 0.965 for HFM on average) between base tiling factor and all blind metrics analysed here. Increasing the base tiling factor increases the overall brightness, the dynamic range and the contrast of the image. Visually however, using a base tiling factor setting which is exaggeratedly high for the target resolution introduces dense pattern artefacts that may deteriorate visual appeal even as they increase the contrast in the image. Changes in the base tiling factor also have an increased effect as more layers are used.

On the other hand, Fig. 13 indicates a strong negative correlation (−0.991 for $\bar{\text{I}}$, −0.991 for DR and −0.975 for HFM on average) between number of layers and all blind metrics analysed here. This can be explained by the fact that the blending algorithm tends to average colours across layers, meaning that using too many layers not only increases processing time but also reduces the range of expressed colours as well as contrast.

Figure 14 indicates a strong negative correlation between base tiling factor and both reference metrics (−0.928 for PSNR [14] and −0.983 for SSIM on average). This may be the numerical consequence of the visual issue described above, where using too many tiles for the target resolution—albeit increasing contrast of the image—introduces strong patterns that may not be relevant to the structure of the map elements being rendered.

[14] The data for 3 layers was discarded when calculating this value as it seems to be an outlier.

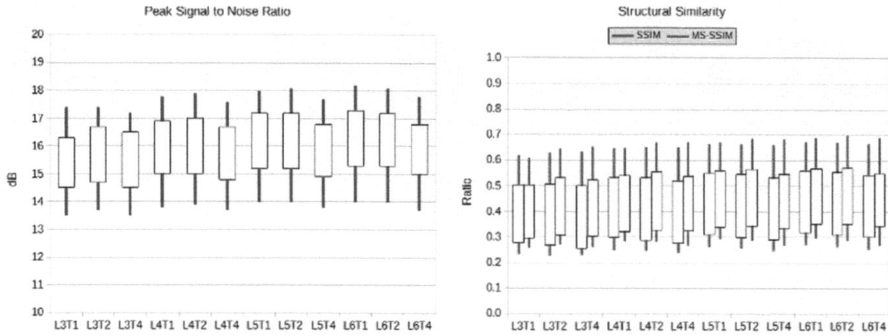

Fig. 14 PSNR and SSIM values for images processed with different settings. *L* stands for number of layers; *T* stands for tiling factor. Whiskers represent minimum and maximum, box represents first and third quartiles for each category.

Table 2 Average PSNR and SSIM metrics for images generated at different settings

PSNR/SSIM	BTF 1	BTF 2	BTF 4
3 layers	15.385/0.408	15.639/0.408	15.452/0.403
4 layers	15.877/0.431	15.943/0.428	15.682/0.419
5 layers	16.096/0.445	16.079/0.438	15.779/0.429
6 layers	16.201/0.452	16.127/0.446	15.816/0.437

Both PSNR and SSIM seem to suggest that the combination of parameters that gives best results is 6 layers starting with a single tile for the whole image as the first layer

Figure 14 also indicates a strong positive correlation between number of layers and both reference metrics (0.943 for PSNR and 0.978 for SSIM on average). This suggests that the layer-blending algorithm, although it tends to smooth out contrast in the image, seems to successfully contribute towards structural detail even at higher resolutions than the generative network was intended for.

5.2 *Processing Times*

The time taken by the system to generate an image is proportional to the number of tiles it needs to texture (which depends on both base tiling factor and number of layers), although the relationship is not exactly linear since more computationally intensive settings tend to make better use of available computational resources with batch processing.

Part of the image generation time is blending, which depends entirely on the number of layers and the target resolution.

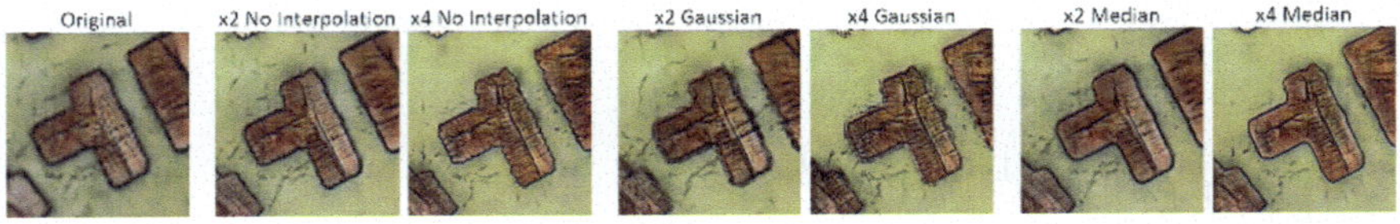

Fig. 15 Impact of zooming in on an input image with different interpolation methods.

5.3 *Image Scaling Potential*

Experimentations were carried out on using the system described above to generate output at a higher resolution than the input image. This is especially important when the input segmentation map is not available in higher resolutions and therefore supporting this feature may help avoid having to redesign or manually remaster the input in such cases.

Figure 15 shows how the system handles upscaling of the input image at a factor of $\times 2$ and $\times 4$. Without any interpolation, this creates stepping (pixelation) artefacts along the edges of objects upon texturing, which is a consequence of the upscaling.

Using any interpolation (such as linear or cubic) or Gaussian blurring a non-interpolated image to get rid of pixelated edges causes artefacts characterised by wide, irregular, double borders around elements as a result of new colour gradients introduced by these methods which the network was not trained upon.

The most visually appealing solution to the above issue can be obtained by Median blurring a non-interpolated image. Both of these methods retain original colours only, ensuring sharp, consistent edges for scaling up to a factor of 4, as shown in Fig. 15. However, Median blurring does cause some rounding of corners, which may be inappropriate for regular shapes typical of buildings when scaling by large factors.

6 Discussion

This being a research carried out on an original context, there are no other works in the same domain to compare with. However, PSNR values for generated images that are generally considered as 'good quality' by modern standards for generative networks typically lie in ranges of 19.67 to 27.39 [31] (video frame conditional synthesis), 6.63 to 14.31 [17] (image extension), 15.264 to 15.971 [16] (panorama synthesis), and 25.11 to 26.77 [18], although—as can be observed—this depends significantly on the task and dataset in question.

Similarly, typical SSIM values for generated images considered as 'good quality' lie in the ranges of 0.66 to 0.88 [31] (video frame conditional synthesis), 0.3951 to 0.4528 [16] (panorama synthesis) and 0.6889 to 0.7369 [34] (fashion item thumbnail synthesis).

In this light, the average PSNR obtained by this system is around 15.8 (min 13.5 at Lr3 Btf1, max 18.2 at Ly6 Btf1) and the average SSIM is around 0.429 (min 0.226

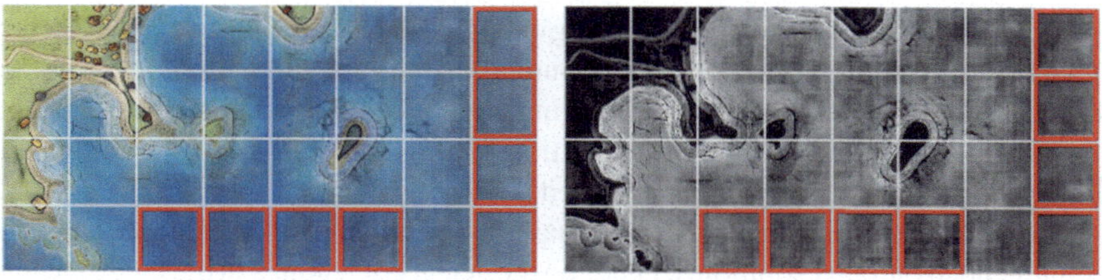

Fig. 16 Some remnant tiling issues are evident in areas with the same feature label. The blending algorithm is not sufficient to smooth out inconsistencies in colour between tiles originating from the same colour label (red areas). The image on the right shows the same detail with increased contrast for ease of observation. No such artefacts are visible on the much more varied land in the same image

at Lr3 Btf2, max 0.672 at Ly6 Btf1). These are overall values across all reference images and all settings. Average PSNRs and SSIMs per setting are shown in Table 2.

The poor performance of the proposed system on the segmentation model indicates that the texturing network combined with the blending algorithm was not capable of imitating the drawing style of the artists whose maps were chosen for the dataset. This does not necessarily mean that the images are not of sufficient quality or that the map elements are not recognisable in the system's output.

Furthermore, given that tiling could not be used to augment the dataset for the segmentation model as it were for the texturing model, the former might have suffered more from the small number of hand-labelled maps than the latter.

Although segmentation analysis seems to indicate that the content generated by the proposed system is not very similar to the dataset, subjective analysis is encouraging with an above-average mean opinion score (3.5/5) and evidence that, in 73% of the cases, output from the proposed system is confused with images from other sources. This leads to the conclusion that the proposed system generates its own style that—even if different from the ones featured in the dataset—is of acceptable appeal that often matches the quality of hand-drawn maps and paid mapping software.

Analysis on the parameters of this system indicates that this system performs best with 5 or 6 layers, and a base tiling factor that corresponds to one-half or the full size of the target resolution. Using smaller tiling factors reduces artefacts but also reduces overall visual appeal, structural detail and contrast. Using fewer layers allows for the use of the full expressive colour range of the network and also improves sharpness but may not heal all seams fully.

The system is generally successful in reproducing high-resolution details and mending seams when there is enough detail in the input segmentation map. In contrast, if a tile happens to contain a single class label only (for example, the tiles marked in red in Fig. 16, which consist entirely of water), the network tends to fail in reproducing a consistent colour. Thus, even with the blending algorithm, differently coloured squares become evident in large areas of the same feature.

This limitation is augmented when using more layers and a larger base tiling factor, given that the more tiles the image is subdivided into, the more likely it is that individual tiles are so small in relation to the full image as to contain one feature only.

For input segmentation maps containing enough detail, the proposed system seems to successfully address the problem of output resolution being determined by the architecture of the network, without leaving any perceptible artefacts. In addition, this system significantly reduces training time for high-resolution models, although slight post-processing times are introduced.

The system described in this paper has no defined maximum resolution. The recommended output resolution depends on the density of features in the input and the scale of the map. Low-density, zoomed-in segmentation maps result in poor appeal for output above 1k resolution (2×2 times the size of the network) as features tend to be scaled up too much during the processing that they are no longer represented in the dataset. On the other hand, experiments on detailed, manually designed segmentation maps of large towns maintained a visual quality comparable to that of images shown to survey respondents even when rendered at 4k resolution (Fig. 19).

7 Conclusion and Future Work

This work has presented a novel approach using which the output of generative adversarial networks can be rendered multiple times and blended to generate output at a resolution that goes beyond the native architecture resolution. The flexibility of the system was demonstrated on multiple sources of input segmentation masks, including manually designed maps, fantasy map generators and real world maps. The impact of the system parameters on the output was analysed in terms of various image quality metrics.

Further work along these lines will include a pyramidal architecture for the generator that guides training at multiple levels, as was done in previous works on generative adversarial networks, for example [22, 23].

It may also be possible to develop and AI-based solution for seamless tiling instead of the algorithmic method described here. In fact, early tests on this work experimented with using inpainting techniques to predict the areas at the seams. This idea was however dropped given that the purpose of the research was to make an resolution-independent model, and using a tiling algorithm for inpainting may solve the seams at the edges of the texturing model output only to create new seams at the edges of the inpainting model output instead.

Future work will also include selecting and segmenting more images for the dataset that was used in this work, given that the small number of images may have impacted the results obtained here. A more complete dataset may be released for public use to provide a common ground for other researchers to compare methods and results with.

The discrepancy between subjective results and segmentation results may indicate the need for better metrics to evaluate artistic output. Current metrics are too dependent on numerical variations between generated and reference images, and poor scores may not necessarily indicate that an image with a differing style has a lower appeal.

This work and future improvements to it may find good use in mapping communities. The proposed method can easily be readapted to other domains, such as modern, sci-fi or post-apocalyptic scenarios, as suggested by early feedback acquired from user opinion surveys.

8 Appendix

See Figs. 17, 18 and 19.

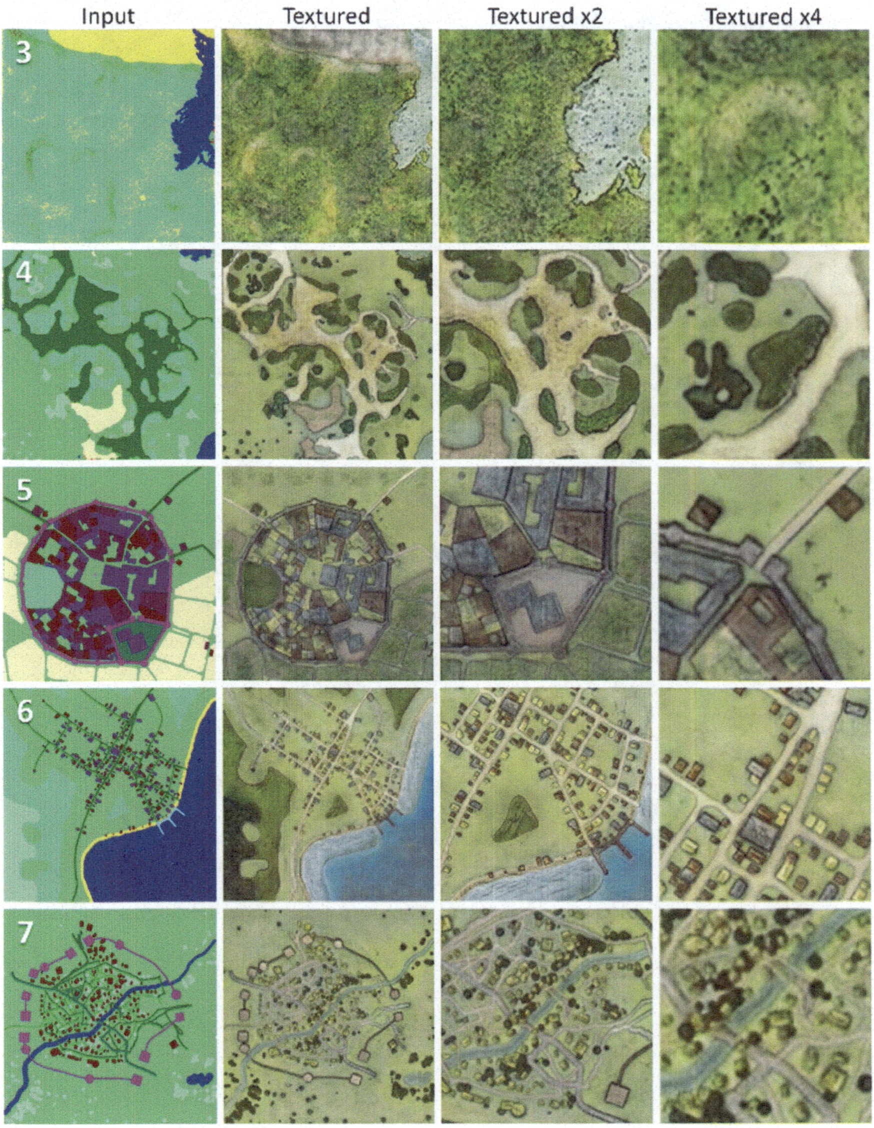

Fig. 17 A set of system output from procedural sources. The numbers correspond to the categories listed in Sect. 4.1

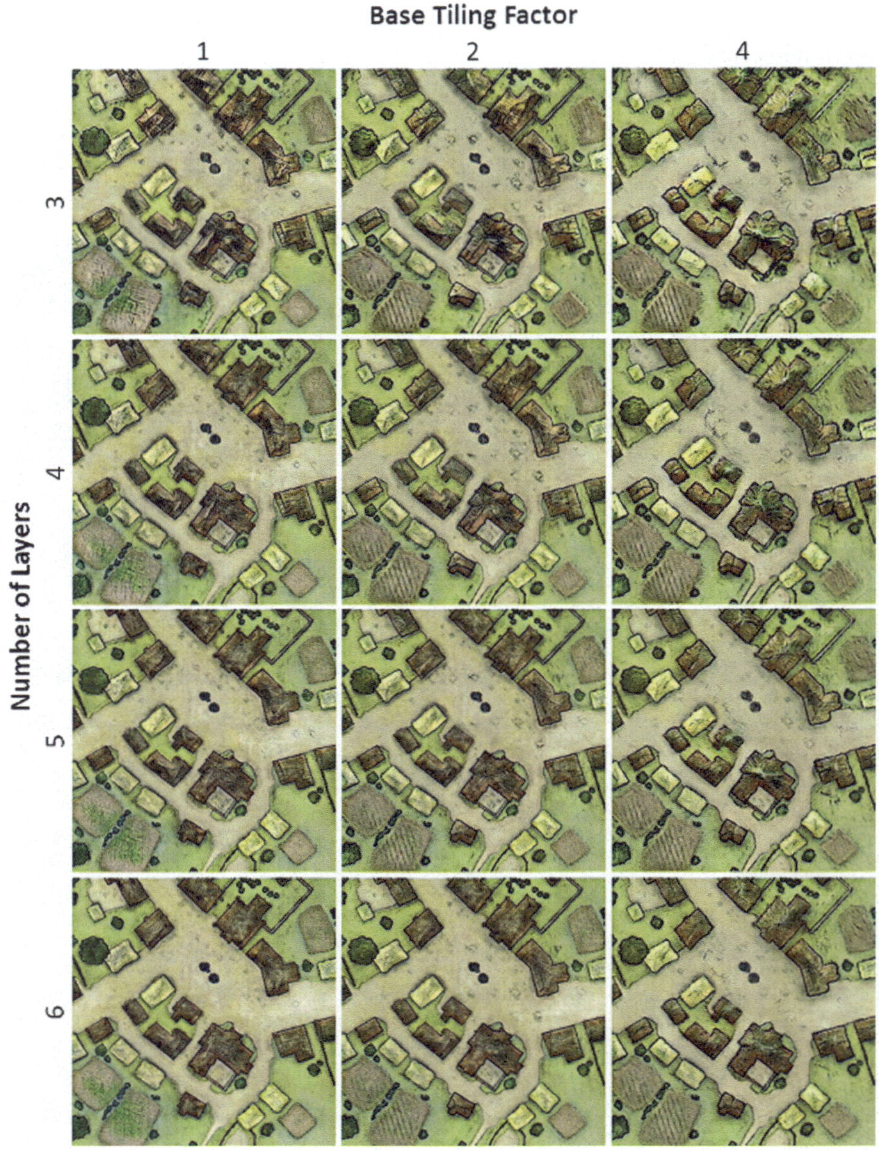

Fig. 18 One of the full-size dataset maps processed with different settings. Original image (not shown here) is an unnamed map by ashlerb

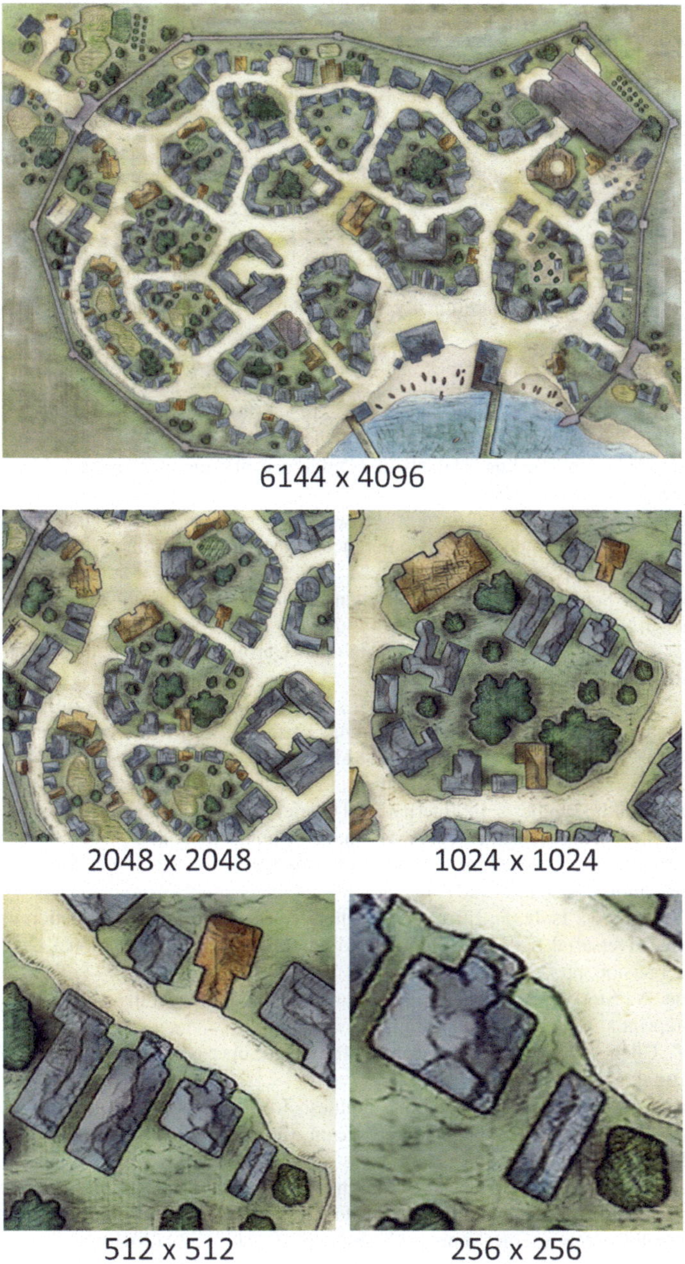

Fig. 19 One of the full-size dataset maps processed at high resolution. This image shows the potential of the proposed system in handling output resolutions that are way larger than the network resolution. The full rendered image is shown along with four consecutive zoomed shots. The image was processed with 4 layers and a base tiling factor of 4 (half the target resolution on the smaller dimension). Original image (not shown here) is that of Rochdale by Venatus Maps. The size of the original image is roughly 2750×1550, demonstrating that the system can also be used to avoid scaling artefacts when upscaling as described in Sect. 5.3

References

1. Siracusa, G., Seychell, D., Bugeja, M.: Blending output from generative adversarial networks to texture high-resolution 2D town maps for roleplaying games. In: 3rd IEEE Conference on Games (CoG) (2021)
2. Goodfellow, I., Pouget-Abadie, J., Mirza, M., Xu, B., Warde-Farley, D., Ozair, S., Courville, A., Bengio, Y.: Generative adversarial nets. In: Advances in neural information processing systems, pp. 2672–2680 (2014)
3. Karras, T., Aila, T., Laine, S., Lehtinen, J.: Progressive Growing of GANs for Improved Quality, Stability, and Variation. International Conference on Learning Representations (2018)
4. Mirza, M, Osindero, S.: Conditional generative adversarial nets. arXiv preprint arXiv:1411 (2014)
5. Chen, X., Duan, Y., Houthooft, R., Schulman, J., Sutskever, I., Abbeel, P.: Infogan: Interpretable representation learning by information maximizing generative adversarial nets. In: Advances in neural information processing systems, pp. 2172–2180 (2016)
6. Park, T., Liu, M.Y., Wang, T.C., Zhu, J.Y.: Semantic image synthesis with spatially-adaptive normalization. In: Proceedings of the IEEE Conference on Computer Vision and Pattern Recognition, pp. 2337–2346 (2019)
7. Arjovsky, M., Chintala, S., Bottou, L.: Wasserstein generative adversarial networks. In: Proceedings of the 34th International Conference on Machine Learning, vol. 70, pp. 214–223 (2017)
8. Berthelot, D., Schumm, T., Metz, L.: Began: Boundary equilibrium generative adversarial networks. arXiv preprint arXiv:1703.10717 (2017)
9. Metz, L., Poole, B., Pfau, D., Sohl-Dickstein, J.: Unrolled generative adversarial networks. arXiv preprint arXiv:1611.02163 (2016)
10. Karras, T., Laine, S., Aittala, M., Hellsten, J., Lehtinen, J., Aila, T.: Analyzing and improving the image quality of stylegan. In: Proceedings of the IEEE/CVF Conference on Computer Vision and Pattern Recognition, pp. 8110–8119 (2020)
11. Engel, J., Agrawal, K.K., Chen, S., Gulrajani, I., Donahue, C., Roberts, A.: GANSynth: adversarial neural audio synthesis. In: International Conference on Learning Representations (2018)
12. Isola, P., Zhu, J.Y., Zhou, T., Efros, A.A.: Image-to-image translation with conditional adversarial networks. In: Proceedings of the IEEE Conference on Computer Vision and Pattern Recognition, pp. 1125–1134 (2017)
13. Zhu, J.Y., Park, T., Isola, P., Efros, A.A.: Unpaired image-to-image translation using cycle-consistent adversarial networks. In: Proceedings of the IEEE International Conference on Computer Vision, pp. 2223–2232 (2017)
14. Antoniou, A., Storkey, A., Edwards, H.: Data augmentation generative adversarial networks. arXiv preprint arXiv:1711.04340 (2017)
15. Wang, Z., Chen, J., Hoi, S.C.: Deep learning for image super-resolution: a survey. IEEE Trans. Pattern Anal. Mach. Intell. (2020)
16. Sumantri, J.S., Park, I.K.: 360 Panorama synthesis from a sparse set of images with unknown field of view. In: The IEEE Winter Conference on Applications of Computer Vision, pp. 2386–2395 (2020)
17. Teterwak, P., Sarna, A., Krishnan, D., Maschinot, A., Belanger, D., Liu, C., Freeman, W.T.: Boundless: generative adversarial networks for image extension. In: Proceedings of the IEEE International Conference on Computer Vision, pp. 10521–10530 (2019)
18. Blanch, M.G., Mrak, M., Smeaton, A.F., O'Connor, N.E.: End-to-end conditional GAN-based architectures for image colourisation. In: 2019 IEEE 21st International Workshop on Multimedia Signal Processing (MMSP), pp. 1–6 (2019)
19. Kim, S.W., Zhou, Y., Philion, J., Torralba, A., Fidler, S.: Learning to simulate dynamic environments with GameGAN. In: Proceedings of the IEEE/CVF Conference on Computer Vision and Pattern Recognition, pp. 1231–1240 (2020)

20. Karras, T., Laine, S., Aila, T.: A style-based generator architecture for generative adversarial networks. In: Proceedings of the IEEE Conference on Computer Vision and Pattern Recognition, pp. 4401–4410 (2019)
21. Salimans, T., Goodfellow, I., Zaremba, W., Cheung, V., Radford, A., Chen, X.: Improved techniques for training GANs. In: Advances in Neural Information Processing Systems, pp. 2234–2242 (2016)
22. Wang, T.C., Liu, M.Y., Zhu, J.Y., Tao, A., Kautz, J., Catanzaro, B.: High-resolution image synthesis and semantic manipulation with conditional GANs. In: Proceedings of the IEEE Conference on Computer Vision and Pattern Recognition, pp. 8798–8807 (2018)
23. Karacan, L., Akata, Z., Erdem, A., Erdem, E.: Manipulating Attributes of Natural Scenes via Hallucination. ACM Transactions on Graphics (TOG) **39**, 1–7 (2019)
24. Jetchev, N., Bergmann, U., Vollgraf, R.: Texture synthesis with spatial generative adversarial networks. arXiv preprint arXiv:1611.08207 (2016)
25. Kim, S., Kim, D., Choi, S.: Citycraft: 3D virtual city creation from a single image. Visu. Comput. **36**, 911–924 (2020)
26. Zhang, J., Wang, C., Li, C., Qin, H.: Example-based rapid generation of vegetation on terrain via CNN-based distribution learning. Vis. Comput. **35**, 1181–1191 (2019)
27. Spick, R.J., Cowling, P., Walker, J.A.: Procedural generation using spatial GANs for region-specific learning of elevation data. In: 1st IEEE Conference on Games (CoG), pp. 1–8 (2019)
28. Wu, H., Zheng, S., Zhang, J., Huang, K.: GP-GAN: towards realistic high-resolution image blending. In: Proceedings of the 27th ACM International Conference on Multimedia, pp. 2487–2495 (2019)
29. Wang, Z., Bovik, A.C., Lu, L.: Why is image quality assessment so difficult? In: IEEE International Conference on Acoustics, Speech, and Signal Processing, vol. 4, pp. 3313–3316 (2002)
30. Narayanam, R., Kumar, K.K.: A novel experimental evaluation for the "development of a novel standard notionâŁž, image quality assessment (IQA) measures, computational time for image contrast. Int. J. Eng. Sci. Res. Technol. **5**, 882–894 (2016)
31. Wang, Y., Bilinski, P., Bremond, F., Dantcheva, A.: ImaGINator: Conditional spatio-temporal GAN for video generation. In: The IEEE Winter Conference on Applications of Computer Vision, pp. 1160–1169 (2020)
32. Xiong, W., Yu, J., Lin, Z., Yang, J., Lu, X., Barnes, C., Luo, J.: Foreground-aware image inpainting. In: Proceedings of the IEEE Conference on Computer Vision and Recognition, pp. 5840–5848 (2019)
33. Wang, Z., Bovik, A.C., Sheikh, H.R., Simoncelli, E.P.: Image quality assessment: from error visibility to structural similarity. IEEE Trans. Image Processing **13**, 600–612 (2004)
34. Pandey, N., Savakis, A.: Poly-GAN: Multi-conditioned GAN for fashion synthesis. arXiv preprint arXiv:1909.02165 (2019)
35. Wang, Z., Simoncelli, E.P., Bovik, A.C.: Multiscale structural similarity for image quality assessment. In: The Thirty-Seventh Asilomar Conference on Signals, Systems and Computers, volume 2, pp. 1398–1402 (2003)

Author Index

A. Dingli et al. (eds.), *Disruptive Technologies in Media, Arts and Design*, Lecture Notes in Networks and Systems 382, https://doi.org/10.1007/978-3-030-93780-5

The manufacturer's authorised representative in the EU is Springer Nature Customer Service Centre GmbH, Europaplatz 3, 69115 Heidelberg, Germany. If you have any concerns regarding our products, please contact ProductSafety@springernature.com

Printed and bound by CPI Group (UK) Ltd, Croydon, CR0 4YY
22/09/2026
02220981-0002